Textile Progress

June 2011
Vol 43   No 2

Assessment of key issues in the coloration of polyester material

T0144595

**Renzo Shamey
and Woo Sub Shim**

The Textile Institute

Taylor & Francis
Taylor & Francis

SUBSCRIPTION INFORMATION

Textile Progress (USPS Permit Number pending), Print ISSN 0040-5167, Online ISSN 1754-2278, Volume 43, 2011.

Textile Progress (www.tandf.co.uk/journals/TTPR) is a peer-reviewed journal published quarterly in March, June, September and December by Taylor & Francis, 4 Park Square, Milton Park, Abingdon, Oxon, OX14 4RN, UK on behalf of The Textile Institute.

Institutional Subscription Rate (print and online): $450/£237/€359
Institutional Subscription Rate (online-only): $405/£213/€323 (plus tax where applicable)
Individual members of **The Textile Institute** can subscribe to *Textile Progress* for just **£40**. Please visit http://www.tandf.co.uk/journals/offer/ttpr-so.asp for further information or contact Jacqui Tearle, Customer Services, Taylor & Francis Group, 4 Park Square, Milton Park, Abingdon, OX14 4RN, UK. Email: jacqui.tearle@tandf.co.uk.

Taylor & Francis has a flexible approach to subscriptions enabling us to match individual libraries' requirements. This journal is available via a traditional institutional subscription (either print with free online access, or online-only at a discount) or as part of the Engineering, Computing and Technology subject package or S&T full text package. For more information on our sales packages please visit www.tandf.co.uk/journals/pdf/salesmodelp.pdf.

All current institutional subscriptions include online access for any number of concurrent users across a local area network to the currently available backfile and articles posted online ahead of publication.

Subscriptions purchased at the personal rate are strictly for personal, non-commercial use only. The reselling of personal subscriptions is prohibited. Personal subscriptions must be purchased with a personal cheque or credit card. Proof of personal status may be requested.

Ordering Information: Please contact your local Customer Service Department to take out a subscription to the Journal: **India**: Universal Subscription Agency Pvt. Ltd, 101–102 Community Centre, Malviya Nagar Extn, Post Bag No. 8, Saket, New Delhi 110017. **USA, Canada and Mexico**: Taylor & Francis, 325 Chestnut Street, 8th Floor, Philadelphia, PA 19106, USA. Tel: +1 800 354 1420 or +1 215 625 8900; fax: +1 215 625 8914; email: customerservice@taylorandfrancis.com. **UK and all other territories**: T&F Customer Services, Informa Plc., Sheepen Place, Colchester, Essex, CO3 3LP, UK. Tel: +44 (0)20 7017 5544; fax: +44 (0)20 7017 5198; email: subscriptions@tandf.co.uk.

Dollar rates apply to all subscribers outside Europe. Euro rates apply to all subscribers in Europe, except the UK and the Republic of Ireland where the pound sterling price applies. If you are unsure which rate applies to you please contact Customer Services in the UK. All subscriptions are payable in advance and all rates include postage. Journals are sent by air to the USA, Canada, Mexico, India, Japan and Australasia. Subscriptions are entered on an annual basis, i.e. January to December. Payment may be made by sterling cheque, dollar cheque, euro cheque, international money order, National Giro or credit cards (Amex, Visa and Mastercard).

Back Issues: Taylor & Francis retains a three year back issue stock of journals. Older volumes are held by our official stockists to whom all orders and enquiries should be addressed:
Periodicals Service Company, 11 Main Street, Germantown, NY 12526, USA. Tel: +1 518 537 4700; fax: +1 518 537 5899; email: psc@periodicals.com.

The 2011 US Institutional subscription price is $450. Periodical postage paid at Jamaica, NY and additional mailing offices. **US Postmaster:** Send address changes to TTPR, c/o Odyssey Press, Inc., PO Box 7307, Gonic NH 03839, Address Service Requested.

Subscription records are maintained at Taylor & Francis Group, 4 Park Square, Milton Park, Abingdon, OX14 4RN, United Kingdom.

For more information on Taylor & Francis' journal publishing programme, please visit our website: www.tandf.co.uk/journals.

CONTENTS

Textile Progress
Vol. 43, No. 2, June 2011, 97–153

Assessment of key issues in the coloration of polyester material

Renzo Shamey* and Woo Sub Shim

Textile Engineering Chemistry and Science Department, North Carolina State University, Raleigh, NC 27695-8301, USA

(*Received 13 February 2011; revised 16 February 2011*)

In a previous publication we reviewed some of the most critical issues that affect the coloration and properties of cotton-based textiles [R. Shamey and T. Hussain, Textile Progress 37(1/2) (2005) pp. 1–84]. Today, polyester is still widely regarded as an inexpensive and uncomfortable fiber, but this image is slowly beginning to fade with the emergence of polyester luxury fibers. Polyester fibers currently comprise a commanding 77% share of the total worldwide production of the major synthetic fibers [F. Ayfi, *2003–2004 Handbook of Statistics on Man-Made/Synthetic Fibre/Yarn Industry. Part One, Fibre for Better Living*, Association of Synthetic Fibre Industry, Mumbai, India, 2004, p. 177]. More than 95% of all polyester fibers manufactured today is based on polyethylene terephthalate. The dyeing properties of polyester fibers are strongly influenced by many of the processing conditions to which each fiber may be subjected during its manufacturing or in subsequent handling. Significant differences in properties of fibers can therefore arise due to their different processing history. Often, the root cause(s) of a problem in the dyed synthetic material can be traced as far back as the manufacturing process. In order to resolve many of the outstanding issues that commonly occur in the dyeing of this important fiber, a comprehensive review of the issues dealing with the manufacturing history as well as fiber processing conditions, including preparation, dyeing, and finishing is warranted. Although some of the underlying problems are related to common causes such as water quality and imperfections in machinery employed, others are specific to the treatment conditions of the fiber. Such conditions include preparation of ingredients, polymerization, fiber and filament processing conditions, as well as heat setting that can cause problems in the coloration of fiber. This summary analysis complements the rich pool of knowledge in this domain and addresses problems in the dyeing of polyester textile materials in various forms. An overview of various textile operations for polyester is given in the beginning. Then, various key steps and critical factors involved in the production of dyed polyester textile materials are described in detail and problems originating at each stage are summarized.

Keywords: polyester dyeing; dyes; colorants; troubleshooting dyeing; polyester pretreatment; polyester polymerization; polyester processing

1. Introduction

Polyester fibers are often regarded as cheap and uncomfortable, but this image is slowly beginning to fade with the emergence of luxury polyester products. Figure 1 shows that polyester fibers currently comprise a commanding 77% share of the total worldwide production of major synthetic fibers [1]. More than 95% of all polyester fibers manufactured today is based on polyethylene terephthalate (PET) [2,3], see http://inventors.about.

*Corresponding author. Email: rshamey@ncsu.edu

ISSN 0040-5167 print / ISSN 1754-2278 online
© 2011 The Textile Institute
DOI: 10.1080/00405167.2011.565151
http://www.informaworld.com

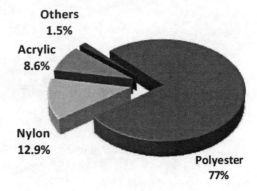

Figure 1.   World production of major synthetic fibres in 2006.

com/library/inventors/blpolyester.htm. The dyeing properties of polyester are strongly influenced by many of the processing conditions to which each fiber may be subjected during
manufacturing or in subsequent handling. Significant differences in properties of fibers can
therefore arise due to their different processing history [4]. Common problems that occur
in the coloration of polyester in various forms are discussed in the following sections.

2. Overview of polyester

British chemists, John Rex Whinfield and James Tennant Dickson, employees of the Calico
Printers Association of Manchester, patented 'polyethylene terephthalate' (also called PET
or PETE) in 1941, after advancing the early research of Carothers [5,6]. They indicated that
Carothers's research did not include the formation of polyester from ethylene glycol and
terephthalic acid. PET is the base polymer used in the manufacturing of synthetic fibers such
as polyester, dacron, and terylene. Polyester is commonly manufactured from petroleum-
based chemical compounds to form fibers, films, and plastics. The starting compounds
[7–9] used in the formation of PET are shown in Figure 2.

Polyester fibers are highly crystalline, mechanically tough, and hydrophobic [4].
Poly(ethylene terephthalate) is generally made from either purified terephthalic acid (PTA)
or dimethyl terephthalate (DMT) together with ethylene glycol. Early polyester production

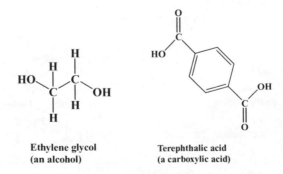

Figure 2.   Starting compounds for the manufacture of polyethylene terephthalate (PET) fiber. Left:
Ethylene glycol (an alcohol); right: terephthalic acid (a carboxylic acid).

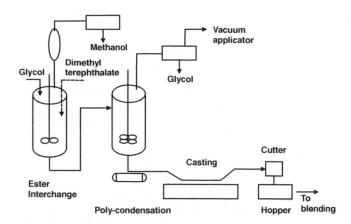

Figure 3. Batchwise production of poly(ethylene terephthalate) from dimethyl terephthalate (DMT).

was based solely on the ester, but with the advent of PTA on a commercial scale in 1964, it became possible to implement processes based on the acid [10]. In general, there are two routes to form polymers: (1) batchwise, and (2) continuous. Early production was entirely batchwise, but the advantages of continuous production, particularly when integrated with the spinning processes, are such that most large-scale production is now continuous. Batchwise productions are now confined mostly to fine filament yarns, variants, and non-fiber end uses.

Batchwise systems usually comprise two reaction vessels: (1) an ester interchange vessel or esterfier, and (2) a polymerization vessel. An ester interchange vessel is equipped with heating facilities (external, internal, or both), an agitator, ancillary vessels for metering in ethylene glycol and molten DMT and for introducing a catalyst, and a fractionating column from which the distillate, mainly methanol, is collected in a further vessel and sent for purification. Esterfiers are also heated vessels, and contain a mechanism for the agitation of the contents. Terephthalic acid may be charged as a solid, or more probably as dispersion in the glycol, along with any additives, from a mixing vessel. The water produced is removed by way of a column to return glycol to the batch and a pressure control valve. A simplified arrangement [7] for the manufacture of the polymer is shown in Figure 3.

Continuous polymerization units usually consist of at least three vessels: (1) an ester interchange or a direct esterification unit, (2) a unit for reducing the excess glycol content and producing a pre-polyester, and (3) a unit for completing the polymerization process. The first stage consists of a stirred vessel or vessels where excess glycol is removed and the molecular weight of the product from the first stage is raised to an intermediate value under slightly reduced pressure. The third stage consists of a reactor or reactors that give a high-surface-to-volume ratio and frequent renewal of surface. The product is extracted from a final reactor by a screw that conveys it either directly to melt-spinning or to a polymer extrusion and chipping unit. A simplified schematic arrangement [8–11] of a continuous polymer production technique is shown in Figure 4 [7].

2.1. Overview of manufacturing of fiber, yarn, and fabric

Typical steps in the production of polyester in the fiber, yarn, and fabric forms are summarized in Figure 5 [12,13]. In the case of direct-spun polymer from a continuous polymerization line, the polymer is supplied in molten form and the polymerization process is

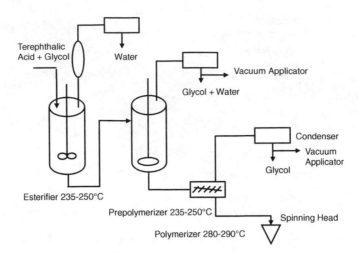

Figure 4.   Continuous production of poly(ethylene terephthalate) from terephthalic acid.

controlled in such a way to produce a uniform feedstock. When the polymer is in solid form, particularly if it is manufactured batchwise, additional processes of blending, drying, and melting are necessary before processing as textile fiber. Polymer characterization and blending may also be needed to ensure uniform feed to spinning, particularly for filament yarn production. Polymer designed for staple fiber does not require such tight control because of the fiber blending processes at a later stage.

Melt spinning requires no chemical reactions and no solvent recovery system. Polymer melt spinning processes fall into the following four classes: (1) relatively low spinning speed, (2) medium spinning speed, (3) high spinning speed, and (4) ultra high spinning speed systems [13–16]. The polymer is dried before melting to prevent hydrolysis. Drying is usually carried out at a temperature of approximately 170°C by passing hot air of controlled humidity through a bed of polymer granules. A very high temperature leads to oxidative reactions and discolorations, whereas a very low temperature leads to insufficient drying [12]. Dried polymer picks up water very rapidly, so it must be fed directly to an extruder or hopper under dry nitrogen without further exposure to air. Relatively low spinning speed processes are directed toward producing a spun yarn or tow with little or no orientation (LOY), which can be subsequently oriented by applying a high draw ratio.

Intermediate spinning speed processes are directed toward producing a partially oriented spun yarn or tow (POY), which can be subsequently oriented by applying a relatively low draw ratio. High spinning speed processes are directed toward producing a highly oriented spun yarn or tow, which normally will not be subjected to an orientation process. Ultra high spinning speed processes lead to products with exaggerated skin-core differentiation and unusual morphology. The highest and the most uniform orientation is obtained by applying a low spinning speed in the order of 3000–4000 m/min and employing a high draw ratio. Although the tenacity of spun yarn rises steadily when spinning speed is increased to 6000–7000 m/min, a pronounced skin-core structure develops above 7000 m/min, with a highly oriented, crystalline and porous skin, and a much less oriented, crystalline and porous core. A product of this type has the interesting property of dyeability in boiling water without a carrier [17,18]. Spun yarn of low orientation is subjected to a stretching or drawing process to convert it into commercially useful yarn of high orientation. Although polyester yarns can be oriented by drawing at room temperature, the stress required to

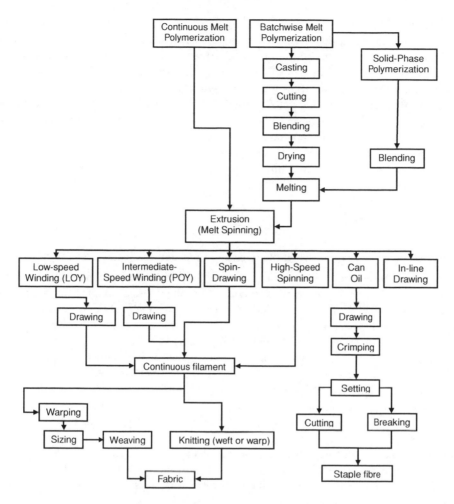

Figure 5. General steps involved in the formation of polyester material in fiber, yarn, and fabric form.

initiate drawing under these conditions is high and the process is found to be unstable. In order to obtain a uniform product, it is necessary to draw at a temperature above the glass temperature of the spun yarn. In practice, the draw temperature is often in the range of 90–100°C. In the case of the filament yarn, the yarn is crystallized through heat setting during the drawing process when it passes over a hot plate at a temperature in the range of 140–220°C, or by using a heated draw roller at a similar temperature. Variations in these processes influence the dyeing property of fibrous polymeric materials, since they affect the relative proportion of the amorphous phase and the chain packing, especially in the intermediate phase between crystalline and amorphous phases [19]. Since the dyeing properties of fiber are affected by the amount of crystalline and amorphous regions, care has to be exercised in setting the drawing ratio.

There are two types of continuous filament fiber: (1) monofilament, which is a single strand of continuous filament, and (2) multifilament, which consists of two or more strands of continuous filament twisted together to form a single strand [20]. Multifilament yarns consist of a bundle of continuous filaments, either twisted or twist-free [21].

2.2. Fabric formation

The manufacturing of fabric is based on two main methods: weaving and knitting. Conventional weaving is carried out by fitting one set of yarns, known as the warp yarns, on the loom to form the length of the fabric. The other set of yarns interlaced at right angles with the warps, are called weft or filling yarns. The yarns can be interlaced in many different ways depending on the weaving patterns used. Common weaves used for most fabrics include plain, twill, and satin patterns [20]. Knitting may be characterized as the inter-looping of one or more set of yarns. Compared with woven fabrics, knitted structures are often more stretchable and therefore specially suited for the manufacturing of undergarments. Prior to processing into apparel and other finished products, however, woven and knitted fabrics pass through several water-intensive wet processing stages. These steps are briefly reviewed in the following sections.

3. Overview of wet processing in the production of polyester material

Polyester may be blended with a variety of natural fibers, the most important of which is cotton. Some of the processes applied to cotton and other natural fibers differ from those employed on synthetic substrates, including polyester. However, a brief overview of these treatments, given below, may provide the reader with a better appreciation of the variables that can influence the outcome of dyeing processes.

The processing of textile fibers is divided into several steps as shown in Figure 6. Grey (greige) fabric may be received from the weaving department in-house or from external sources. After unloading, the greige fabric is piled in the store and inspected. Smaller length fabric pieces are then stitched together to allow for smooth fabric transition through machinery and avoid major production interruptions. Initial inspection includes examining the fabric for the presence of holes, untidiness, and general quality on an inspection table where amendable faults are repaired manually using needles and thread. Nonrepairable fabrics are rejected and sent back to supplier, whereas fabrics with acceptable quality are transferred to appropriate processing units on manual trolleys. This process is known as *reception*.

In general, the first step in the pretreatment of greige fabrics is called *singeing*, which is carried out to remove the protruding hairs from the fabric surface. These protruding fibers can give the cloth a matt and raw appearance even after finishing. Singeing units are generally combined with desizing machines as ancillary components. There are two chambers on a desizing machine. Singeing is often carried out in the first chamber and desizing in the second. In singeing, fabrics are passed over a series of burners, where the loose hairy fibers protruding from the surface are burnt resulting in a smooth and clean fabric appearance.

The process of *desizing* mainly removes the sizing material such as starch, wax, and oil that are applied to yarns prior to weaving to ease production and reduce yarn breakage. *Scouring* is a cleaning process in which oil- or wax-based impurities are removed. After scouring fabric becomes more absorbent compared to its greige counterpart; however, the natural coloring matters present in natural fibers are often not removed. Such colored impurities are removed through *bleaching* using bleaching agents such as hydrogen peroxide or sodium hypochlorite. The fabric is generally treated in a rope form and chemicals are prepared in separate containers and poured in the machine's trough. After scouring and beaching, fabric is treated with an acid to neutralize the alkalis present to prevent subsequent fabric damage and degradation, as well as skin irritation. *Neutralization* is carried out in an

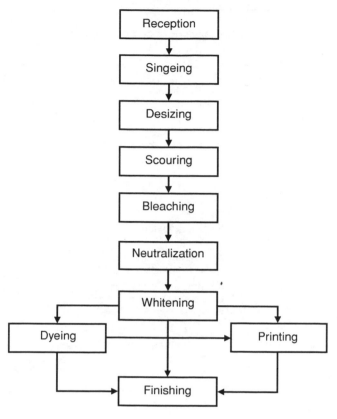

Figure 6.   Major steps in processing of textile fibers.

acidic bath, which is subsequently discharged and the fabric is washed with fresh water to remove traces of acid from its surface.

Generally, polyester fabric is not mercerized although caustification may be carried out on customer demand and based on the required quality of product. In order to improve the whiteness and obtain bright and 'clean' substrates, a whitening process involving the application of optical or fluorescent brightening agents may be carried out. Some printed textiles, especially those dyed to light and pastel colors, are often brightened optically.

While more than one color may be generated via dyeing, the process mainly implies the uniform application of a single color to a textile product. Printing, on the other hand, often implies the application of several colors to specific areas of textile substrates based on a specific pattern. Three main production methods in textile dyeing are batch, semi-continuous, and continuous processes. Batch dyeing involves the application of dyes from a solution or dispersion at specific liquor to goods ratio to textile substrates where the depth of the color obtained is mainly determined by the amount of colorant present in relation to the quantity of fiber, although several other factors also influence the overall dye uptake. Semi-continuous dyeing is characterized by the application of dye(s) in a continuous mode while fixation and washing steps are run discontinuously. Continuous dyeing is operated at constant composition of chemicals in several troughs where a long length of textile fabric is pulled through each section of the continuous production line and includes fixation and wash-off processes.

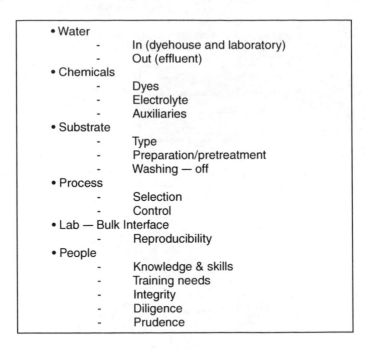

• Water
  - In (dyehouse and laboratory)
  - Out (effluent)
• Chemicals
  - Dyes
  - Electrolyte
  - Auxiliaries
• Substrate
  - Type
  - Preparation/pretreatment
  - Washing — off
• Process
  - Selection
  - Control
• Lab — Bulk Interface
  - Reproducibility
• People
  - Knowledge & skills
  - Training needs
  - Integrity
  - Diligence
  - Prudence

Figure 7. Key factors for improving dyehouse productivity.

In printing, color is applied only to specific sections of the cloth, according to a preset pattern, based on a number of printing techniques that include screen printing and inkjet printing. Dye fixation is carried out by steaming or baking the printed material followed by washing to remove surplus dye and thickeners.

Last but not least, *finishing* operations endow the fabric with a particular appearance, surface texture, or characteristic such as water repellent, stain repellent, antistatic, biocide, stiffened, shape retentive, and wrinkle resistant. These processes can have a profound impact on the final color of the substrate [22–26].

Having reviewed the main operations involved in the production of textile substrates, it is pertinent to discuss key considerations that can improve dye house productivity as well as successful application of color. Some of the main factors that influence the industrial application of color are displayed in Figure 7 [27–32].

The role of some of the main factors in the success of the coloration processes is described in the following sections.

4. Coloration problems originating from manufacturing polyester

4.1. Problems caused by polymerization

The orientation and crystallinity of polymer chains have a significant impact on fiber properties. Highly crystalline and oriented fibers are often stable to changes in temperature and moisture but have relatively low elongation to break, low absorbency, high stiffness, and are generally difficult to dye.

Orientation and crystallinity of fibers are determined during the polymerization step. The diminution of the degree of polymerization and the associated fall in tenacity and extension are largely attributed to a cleavage of macromolecular chains. Polymerization is generally carried out in a closed chemical reactor containing a number of inlets and a product (polymer) outlet. The temperature and pressure inside these reactors are closely controlled.

The dye uptake in polymer is controlled by the number of crystals present and the orientation of the amorphous phase [33,34]. This is based on the fact that the dye diffusion depends on the segmental mobility of the amorphous region, which in turn depends on the regional order as well as the number, size, and size distribution of the crystallites. The porosity of the fiber, i.e. the number, size, and distribution of the recesses in the matrix, can also affect dye uptake [35].

Variation in the concentrations of monomers, temperature, and the time of polymerization can also alter the end-group concentration [36–40]. Side reactions involving the aldehyde group and diethylene glycol (DEG) during the polymerization of polyester fiber also influence the characteristics of the final product. For instance, increasing the concentration of DEG in polyester fiber increases the dyeability of the fiber. However, presence of DEG lowers the melting point of polymer and results in the formation of an inferior polymer with poor strength and high sensitivity to ultraviolet light. In addition, it has been observed that polymerization at relatively high humidity levels results in increased diffusion of dye molecules into the substrate [39–41].

Several approaches have been attempted to improve the dyeability of polyester, a recent example involves subjecting polyester fibers to vinyl graft polymerization using either radiation [36,42,43] or chemical initiation prior to dyeing [44–48]. The color strengths of the grafted samples have been found to be higher than their unmodified counterparts. Grafting with various monomers has been reported to result in different depth of shade of dyed polymer as shown below [42]:

MMA (methylmethacrylate) > styrene/MMA > styrene > MAA (methacrylic acid).

Differences in the color strength obtained with various monomers could be attributed to differences in [45–47,49]

- affinity of the monomer to both dye and substrate;
- molecular size of the monomer;
- hydrophobic/hydrophilic balance of the monomer;
- ability of the monomer for intimate association with and/or dissolution of dye;
- ease of conversion of the monomer to oligomers and molecular weight of the latter;
- ability of the monomer and/or its oligomers to form a film containing dye in the case of polyester; and
- dissolution of oligomers in their own monomer.

However, the presence of monomers could also favorably affect the fiber's swelling ability and diffusion of dye molecules into, and their absorption on, the fiber.

4.2. Problems caused by glass transition temperature

The glass transition temperature (T_g) of a polymer has considerable influence on the way in which a dye diffuses within the polymeric matrix. It has been shown that T_g is influenced by the size and the shape of molecules, by interaction between the dye and the polymeric matrix, and that dyeing can result in a reduction of T_g [50,51]. In the dyeing of polyester, dyes are commonly applied as dispersions in water. Slight solubility of dye in water is advantageous, since it permits diffusion from aqueous medium to the fiber surface. Substantial solubility of dye in water, however, is disadvantageous, since it leads to poor dyebath exhaustions. A consequent challenge posed by the relatively high glass–rubber transition temperature of the fiber is to obtain a sufficient diffusion rate of the dye into the fiber to build up an adequate shade without compromising the fastness properties of the dyed substrate. There are four main solutions to this problem as shown below [52].

4.2.1. The use of carriers

Carriers are organic compounds added to the aqueous dyebath to increase the rate of dye diffusion at the dyeing temperature [23,40]. Carriers act as fiber plasticizers and are preferably of poor solubility in water. Among the more important varieties of carriers are O-phenylphenol, biphenyl, methylnaphthalene, O-dichlorobenzene, and methyl salicylate [53]. These are commonly dispersed in bath at 1–8 g/L together with the dyestuff, migrate relatively rapidly into the fiber, plasticize the noncrystalline regions, and thus reduce the glass transition temperature. Dye diffusion is influenced by the extent of reduction in polymer's T_g due to plasticizing effect of the carrier. The use of carrier permits the application of dye to fiber at atmospheric pressure. The ideal carrier for use in dyeing should produce appreciable color yield at boil in a practical dyeing time at low carrier cost, and without seriously affecting physical properties of fiber or fastness properties of dyes. The application of carriers should not cause spotting or noxious effects because of toxicity during the coloration process.

However, there are disadvantages to this technique. Carriers are expensive enough to add significantly to the cost of dyeing; they are undesirable from an environmental perspective; some are toxic, and their residual presence in the fiber can lead to fiber staining, unpleasant odors, and reduced light fastness. For these reasons, whenever possible, the application of carriers is not favored. If carriers are to be applied, they are preferably removed by treatment of dyed substrate with hot air after dyeing. Carriers must, however, satisfy the following requirements [23,29,30,53–56]:

(1) Must not be toxic or mutagenic.
(2) Must not have too high a vapor pressure.
(3) Must not volatilize in steam.
(4) Must be readily emulsifiable or soluble.
(5) Should be inexpensive.
(6) Must be very effective.
(7) Must be readily removable and not have an adverse effect on dye fastness.
(8) Must cause no degradation or discoloration of the fiber.

Typical carriers accelerating the dyeing of polyester are as follows [29,56–58]:

(1) Aromatic hydrocarbons such as diphenyl, naphthalene, and toluene.
(2) Phenols, e.g. phenol, o- and p-chlorophenols, o- and p-phenylphenol, and m-cresol.
(3) Chlorinated aromatics: mono, di- and tri-chlorobenzene and chlorinated naphthalene.
(4) Aromatic acids, including benzoic, chlorobenzoic, and o-phthalic acids.
(5) Aromatic esters, e.g. methyl benzoate, butyl benzoate, dimethyl and diethylphthalate, DMT, and phenyl salicylate.
(6) Aromatic ethers such as p-naphthylmethyl ether.
(7) Miscellaneous compounds, including acetophenone, phenyl cellosolve, phenylmethyl carbinol, methyl salicylate, and benzanilide.

Some of the disperse dyes recommended for dyeing with carriers include the following [59–61]:

(1) C.I. Disperse Yellow 42, 218, 229, and 236.
(2) C.I. Disperse Orange 13, 25, and 37.
(3) C.I. Disperse Red 13, 50, 60, 73, 91, 92, and 343.
(4) C.I. Disperse Violet 26 and 28.
(5) C.I. Disperse Blue 56.

4.2.2. High-temperature dyeing

Super atmospheric pressure or high-temperature (HT) dyeing is usually carried out at a temperature of 125–130°C and a correspondingly elevated pressure. At such temperatures the diffusion rate of dyes is high enough to produce satisfactory shades in a dyeing time of about 1 h. High-pressure beams, jets, and jigs can be used. Although such equipments tend to be more expensive than conventional atmospheric pressure dyeing vessels, their cost is more than offset by the omission of carriers. High-temperature dyeing involves the following advantages compared with carrier dyeing [18,62]:

- It does not adversely influence the light fastness.
- The potential toxic effect because of the use of carrier dyeing is eliminated.
- No additional effluent problems arise.
- No expenditure on carriers.
- Unlevel dyeings and spotting are obviated.

Unfortunately, the extraction of cyclic trimer from the fiber during high-temperature dyeing can sometimes cause problems. The trimer crystallizes from the dye liquor on cooling, deposits in cooler regions of the vessel, and may contaminate the fabric. Additives are marketed to minimize the formation and quantity of deposits. Processes involving selective hydrolysis of the dissolved trimer by alkali either at the start or at the end of the process have been described elsewhere [62]. Removal of deposited trimers via the hydrolysis technique requires careful selections of dyes since many disperse dyes are degraded under alkaline conditions. Discharging the dye liquor at the dyeing temperature without cooling is a very effective means of reducing the formation of deposits but this method requires a second pressure vessel. It is important that pressurized vessels are regularly cleaned.

4.2.3. Thermofixation

Thermofixation (commercially known as Thermosol® process) is particularly suited to continuous dyeing processes, since it involves padding the dyestuff from dispersion onto the fabric, drying, and then heating to a temperature of 180–220°C. At such temperatures, diffusion rates are so high that a few seconds suffice for adequate penetration of dye molecules into the substrate. A number of variables, however, affect the final shade of the dyed fabric during the thermofixation process. Some of these variables include thermosol period, temperature, type of disperse dyestuff, and pad bath auxiliaries. The advantages of thermofixation can be summarized as follows [19,63–66]:

- Excellent dye utilization.
- No carrier is required, thus the chances of spotting and impairment of light fastness are eliminated.
- The fabric is processed in open width, thereby eliminating the problem of rope marks.

- Dyeability is not affected by previous heat setting.
- The energy consumption is lower than that in a batch process because of shorter dyeing cycles, lower liquor ratio, and use of energy recovery devices.
- Also, a large yardage can be dyed more economically than by batch processes.

Some of the shortcomings of the thermofixation dyeing of polyester are listed below [19,63–66]:

- 'Listing' can occur due to different nip pressure and migration of color during the intermediate drying.
- Ending and tailing marks may form due to high affinity of the dyes.
- Two sidedness may occur due to uneven drying on two sides of the fabric or due to uneven padding.
- Pale spots or specks may be formed due to condensation of water or foam, as well as the residual alkali on the goods.
- Pale shades may be formed due to poor penetration of dyestuff.
- Dark or pale selvedge may be present due to feeding liquor from one side of the padding trough only.
- Frosted dyeing may be caused by color migration during intermediate drying or thermofixation stage.
- Widthways variation in drying may occur due to irregular air jets and velocity of air circulation in stenter.
- Foam may be generated by the excessive use of dispersing agents and other auxiliaries in pad liquor.
- Colored specks may be present due to incorrect concentration of dyestuff and auxiliaries.

The disperse dyes applied to polyester by Thermosol® require a suitable heating time to ensure good penetration of the dye [67,68]. Sometimes, particulate migration in fabric padded with disperse dyes during intermediate drying prior to thermofixation can be a serious problem, which can lead to shade variation and unlevelness in continuous dyeing procedures. Several methods have been suggested to prevent migration as listed below [68–70]:

(1) Uniform preparation of the fabric.
(2) Use of low wet pick-up in padding.
(3) Use of antimigrants in the padding solution.
(4) Use of disperse dyes with low concentration of dispersing agent.
(5) Drying all sections of substrate uniformly.

4.2.4. Modification of the chemical structure of fiber to reduce its glass transition temperature

Modification of the polymer to reduce T_g has been attempted to increase the rate of dyeing of polyester fibers [71,72]. This can be achieved by incorporating a small amount of co-monomer within the fibers' backbone. The most effective co-monomers are aliphatic in character. Replacing a small proportion, usually 5–10 mol%, of terephthaloyl unit by units derived from an aliphatic dicarboxylic acid such as glutaric acid (also called pentanedioic acid: $HO_2C(CH_2)_3CO_2H$) or adipic acid (also called hexanedioic acid: $(CH_2)_4(CO_2H)_2$) produces fibers that will dye at the boil without carriers. Aromatic units, for example, based on isophthalic acid, act primarily through reducing crystallinity, and these are therefore

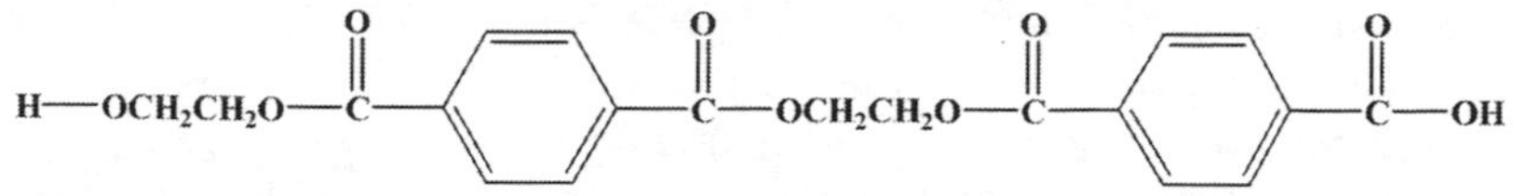

Figure 8. Molecular structure of a linear ethylene terephthalate (ET) dimer.

less effective. Reducing the crystallinity of the oriented polymer reduces T_g and increases the rate of diffusion of dyes. Modification by copolymerization technique is not wholly beneficial; however, many physical properties such as recovery, crease retention, and resistance to shrinkage may be altered [73,74].

4.3. Problems caused by oligomers

Oligomer is a general term given to a polymer with a small chain length usually made up of a few molecules of the basic unit; in the case of polyester, two or more ethylene terephthalates, i.e. linear dimer (Figure 8) or trimer (Figure 9) [75], are produced as a side reaction during the manufacture of the polymer. Approximately 0.1–1.0% oligomer is produced as a side reaction in the manufacture of polyester. Oligomers, as low molecular weight polymers, are released into the dyebath when polyester is dyed. The oligomer, because of its smaller chain length, is in the form of a fine powder that is entrapped in the polymer and during high-temperature polyester dyeing migrates to the surface. These surface deposits form a crystalline structure with a melting point of around 317–320°C and exhibit birefringence in polarized light. The displacement of oligomer to the fiber surface is higher at increased depths of shade. The surface oligomer content almost doubles on dyeing dark and extra dark shades. These oligomers can deposit on the yarn or the surface of the dyeing machinery. Oligomeric deposits on dyeing machinery result in lost productivity because of the time required to clean the machine with alkaline chemicals. When the amount of oligomers increases, it manifests itself in excessive white powder formation on rings and ring rail. Oligomers also cause problems in spinning of dyed fibers. This behavior causes problems during dyeing and finishing stages [48,53,75–77]. Some of these issues are listed below:

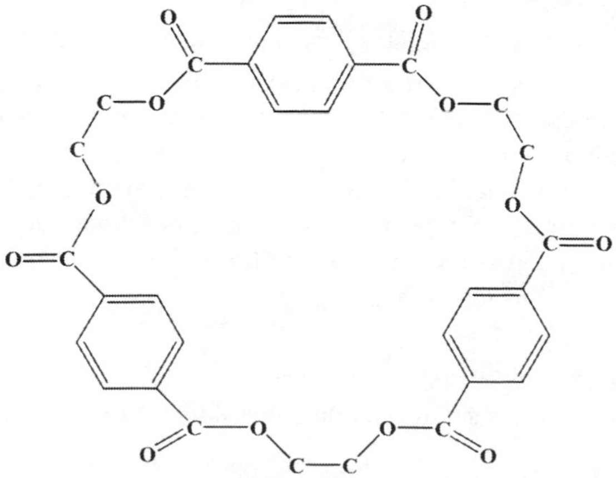

Figure 9. Molecular structure of a cyclic ET trimer.

(1) Oligomeric deposits hinder uniform fabric passage especially when dyeing tightly wound goods/packages, and this increases the risk of unlevelness.
(2) Oligomer deposits on pump result in improper pump pressure.
(3) Oligomers can choke spindle holes that disrupt appropriate flow patterns.
(4) The effectiveness of the pump decreases due to oligomeric deposits in yarn and beam dyeing increasing the risk of unlevel dyeing.
(5) Oligomer deposits on heating elements result in variations in rate of temperature.
(6) Disruptions to heating, fluid flow, and pump circulation also increase overall energy costs of the process.
(7) During the processing of dyed polyester yarn, abrasion causes oligomeric deposits in the machine, which results in considerable dusting of the surroundings.
(8) Oligomer deposits on winding or twisting machinery guides cause high tensions and increase friction on the yarn, leading to poor package construction and end breaks.
(9) The processing properties of yarn (especially sewing threads and in weaving) deteriorate due to high yarn friction coefficients caused by the presence of oligomers, which cause an increased number of broken threads.
(10) More frequent cleaning of dyeing and winding machines is needed due to the formation of oligomeric deposits within machinery, which increases downtime and lowers efficiency.
(11) Dusting during coning occurs, which requires care to prevent potential health hazards.

Also, several unlevel dyeing problems are caused by oligomeric deposits, which are summarized below [48,53,75–77]:

(1) Filtration effect (inside–outside variation) can be caused due to the build-up of oligomers in yarn packages.
(2) Textile fabrics very often tend to 'chalk' due to the presence of powdery oligomer deposits on the surface.
(3) Oligomers decrease the color brilliance of the dyed fiber because of scattering of diffused light, which makes the surface appear dead and dull, especially in dark shades such as black, navy, etc.

The surface oligomer concentrations can be significantly reduced in an alkaline medium. Oligomers can be suspended in the reduction clearing liquor and removed when the liquor is drained. Oligomer deposits on the fiber surface are therefore removed by the reduction-clearing process, although more severe treatments may also be necessary to remove the cyclic trimer from dyeing vessels [78]. Several recommendations have been put forth to resolve problems pertaining to deposition of oligomers, although it has been stated that these solutions do not address the wide range of associated problems [79–81]. Methods employed include the following:

- Use of carriers based on chlorinated benzenes.
- Incorporation of lubricants to facilitate dissolution of oligomers and prevent their redeposition.
- Dropping dyebath at high temperature.
- Application of alkaline reduction clearing after scouring.

4.4. Problems due to melt spinning

The properties of a synthetic fiber are determined by their spinning conditions. Among the melt spinning factors that determine fiber properties, tension acting on the fiber and fiber temperature are important [31]. In addition, the following quality control measures should be considered in the spinning process [13,31,76]:

- Draft distribution and setting.
- Type of ring and traveller.
- Spindle speed.
- Twist.
- Lift.
- Creel type.
- Feed material.
- Machine length.
- Drive type.

Denier is controlled by ensuring that the flow of polymer through each spinneret hole is uniform. However, if a hole is dirty or contains polymeric residues, its effective diameter would be reduced and the extruded filament would become finer. If the spinneret has been used for substantial periods, then some of the holes would be worn out and filaments emerging out would be rough and nonuniform.

The undrawn or partially oriented yarns produced by melt spinning have relatively low degrees of molecular orientation and are easy to dye. Drawn materials, on the other hand, show much higher degrees of molecular orientation [12]. Excessive heating of polyester during the extrusion process, triggered by too high spindle speeds, can result in deformation, flattening, twisting, and partial bonding of fibers to each other [7].

Variations in fiber diameter because of the spinning process cannot be corrected later and may even be exacerbated during the drawing process. Such variations influence the dyeing behavior and the final appearance of the finished goods. A range of factors affecting variations in fiber diameter include [14–16,67,77] the following:

- Spinneret hole diameter.
- Poor spinneret design, crowding of holes.
- Turbulence in the solidification zone, draft in the cooling chimney, variations in the cooling air-flow rates, and inefficient cooling.
- Variations in the coagulation bath concentrations.
- Fluctuations in throughput.
- Presence of gels, naps, cross-linked polymers, polymer dusts in filtered melt-dope that pass through the spinnerets.
- Poor quality polymer melt/dope containing big particles of additives, dulling agents, oxidized or cross-linked polymers and foreign bodies.
- Polymers with different melt rheology, and redried polymer.

4.5. Problems due to drawing and draw line

After extrusion, the fiber is said to be unoriented, undrawn, or as-spun. At the initial manufacturing stage the oriented polymer is largely amorphous. The polymer chains within the fiber begin to orient themselves along the longitudinal axis of the fiber because of the flow of the long polymer chains through the spinneret hole. However, the degree of orientation and crystallinity of fiber at that point is low. Polyester fibers are produced at various draw ratios, ranging from the relatively low values used for some staple fibers and draw-bulked yarns, to the high values for high-tenacity products, some of which may have received further hot-stretching treatments [4]. The drawing process enhances the orientation of molecular chains and causes an increase in close-packing of chains and polymer crystallinity. In general, the diffusion rate of dye increases with decreasing draw ratio [78,79]. When the draw ratio increases, dye uptake decreases because of the changes in the orientation of crystallite and amorphous sectors. While the change in the orientation of crystallites is slight, the orientation of chains in the amorphous region increases substantially. Therefore, increased overall orientation within the fiber decreases the amount of dye uptake. The solidification temperature of amorphous PET lies between 68°C and 69°C, and that of crystalline PET between 79°C and 81°C [22]. For heating temperature below the crystallization temperature, the fiber extension values increase with an increase in temperature while the tenacity diminishes.

Indeed, all major fiber properties, including denier, tenacity, elongation at break, crimp properties, spin finish, shrinkage, and dyeability are shaped in the draw line. Polyester filaments with low preliminary orientation exhibit the largest variation in the cross section in cold drawing. With increasing pre-orientation, the natural draw ratio and variations in cross section are reduced. The shrinkage tendency of polyester filament also generally diminishes with increased drawing rate. The drawing speed exerts almost no effect on polyester filament with a high preliminary orientation.

Two of the most important factors that influence fiber properties in a spinning mill are spin finish and crimp. Spin finish is applied to the undrawn tow at melt spinning stage essentially to provide cohesion and static protection. In the draw line, a major portion of this finish is washed away, and a textile spin finish is put on the tow by either kiss rolls or a spray station. This textile finish consists of two components, one that gives cohesion and lubrication and the other that confers static protection. The amount of spin finish applied is decided by the end use of the fiber. An excessive amount of spin finish can adversely affect dye uptake and levelness properties of dyed fiber.

The second important factor after spin finish is crimp. Crimping is required to ensure smooth running of fibers through various production steps. After the tow is crimped, the crimps are set by passing the tow through a hot air chamber. The temperature at this point can also affect the tenacity and dyeability of the fiber. These properties are also controlled by the draw ratio and annealer temperature, which is usually set at 220°C.

4.6. Problems due to cutting

Cutting defects can be detected during production since smoothly cut fiber ends are only marginally dyed, while squeezed fiber ends are dyed intensely [32]. A highly tensioned tow is first laid over sharp blades and pressed down by a pressure roll and the filaments ends are then cut. Overlength fibers are those whose length is greater than the cut length plus 10 mm and these are often caused by broken filaments, which cannot be straightened by tensioning at the cutter [20,21,80].

The success of the cutting process varies depending on the degree of disturbance of the original fiber arrangement on the converter during the cutting process. In addition, if some blades become blunt, then the pressing of tow onto the blades creates high temperatures and so tip fusion occurs where tips of neighboring fibers stick to each other and form fiber clusters that do not exhibit adequate dye uptake. The separation of fibers in such clusters is almost impossible. However, if these sections are not removed, clusters will be transferred into the yarn and cause a yarn fault. Moreover, if the blades are not very sharp, they will not produce straight edges and some rounding of fiber tips at the cut edge could result. Such fibers are called nail heads and constitute a form of fault.

When fibers are cut they fall down by gravity into the baler. Because of crimping, clusters may be formed. Clusters should be opened out as they can otherwise result in choking in blow-room prior to chute feeds. Fiber clusters are opened by using a ring of nozzles, pointing upwards, below the cutter through which high pressure air jets operate.

4.7. *Problems due to fiber size and cross-sectional shape*

The size of fiber, normally described by denier or tex, as well as their cross-sectional shapes have direct effects on the surface properties of fibers as well as many yarn and fabric properties. Properties that depend on the proportion of available surface areas of fibrous products, such as dyeing and chemical reactions, are expected to vary with fiber size [81]. Although the rate of chemical reactions should be enhanced by higher surface-to-volume ratios, in fibers with smaller diameters, the final results may not be equally positive. More dye is needed to achieve the same shade in dyeing fabrics containing microdenier fibers than those of regular denier fibers. This is because the higher surface-to-volume ratio of smaller fibers enhances dye adsorption and, thus, dye uptake. In addition, the increased curvature of fibers enhances with decreasing fiber diameters, while, on the other hand, increased surface also increases light scattering. The increased light reflectance in smaller fibers because of higher curvatures results in a lighter appearance, for the same amount of dye applied, in comparison to coarser fibers [82]. Consequently, larger amounts of dye are needed to dye fabrics containing microdenier and finer fibers than those with regular denier fibers. In terms of cross section, the study of suitably stained cross sections of polyester fibers, under the electron microscope, show a reticulate pattern with a few small voids that are possibly associated with the presence of solid particles of delustrants [11]. It is reasonable to infer that the changes in dyeing behavior that have been described earlier are associated with changes in the size, spacing, and distribution of these highly ordered regions and that these influence, and are influenced by, the number and arrangement of chain folds in the materials [7,83].

5. Problems originating in the formation of yarns

Several parameters, which include denier variations, periodic denier, filament count, twist per inch/twist direction, fiber cross section, lustre, tensile strength, elongation at break, and broken filaments, affect dyeability of polyester yarns [15,21,84–86].

Polyester is produced as medium and high tenacity staple fibers of various lengths. Staple fibers are usually drawn to give medium tenacity, but may be spun from polymers of relatively low average molecular weight to give improved pilling performance. In addition, staple fibers are generally crimped and many of the filament yarns are bulked by false-twist or some other texturing process [87]. Many complex factors exert an influence on these processes and their inter-relationship is not completely resolved at present [88,89]. Among the factors that influence these processes are polymers melting point, spinning temperature, and viscosity, as well as shape, dimension, and arrangement of the spinneret orifices; in addition to the amount of feed doffing speed and cooling conditions, such as construction of the duct, cooling medium, and temperature.

Polyester fibres are treated with lubricant and antistatic to assist in the spinning and other associated mechanical processes. These are used to help with the production of suitable yarn and fabric for warping, weaving, or knitting. In dyeing of polyester and cotton blends, however, the application of these surfactants may decrease fiber dyeability and moreover they can appear as specks in the dyed material [90,91]. Also, surfactants on the yarn or fabric surface can hinder the diffusion of the dye and stain its surface [62]. Substrates 'contaminated' with surfactants, however, are readily cleaned by comparatively mild scouring procedures involving dilute solutions of soap or anionic, or nonionic surfactants (or their mixture) containing trisodium phosphate or sodium carbonate [29,61,92,93].

Polyester yarns are semi-crystalline in the sense that their density is midway between that of highly organized crystals and the extreme randomness of an amorphous structure [94]. Undrawn or partially oriented yarns produced by melt spinning are relatively easy to dye. Such materials have relatively low degrees of molecular orientation, which explains their rapid dyeing properties. On the other hand, drawn yarns show much higher degrees of molecular orientation and dye at much slower rates. Filament yarns must also be twisted to a level that would permit the production of reasonably permeable dyeing packages and the twist must be set before the yarn is rewound. Flat or very-low-twist continuous filaments used for the production of yarn packages lie very closely together. Such packages have very low permeability and for economically acceptable weights of yarn are difficult to dye uniformly.

Filaments are cut to staple fibers to produce spun yarns that may be similar in touch to natural fibers. As tactile properties of filaments are often described as artificial, they are specially treated or textured to improve their appearance and function. One main advantage of using filaments is that pilling is heavily reduced. Polyester filament yarns may be dyed as wound packages on rigid, perforated conical, or cylindrical centers, on cylindrical interlaced springs or, more rarely, in the form of mock-cakes or hanks [12]. Generally the most satisfactory form of dyeing filament packages is in the form of cheese, cross-wound on a stainless, interlaced spring, preferably of the type that has some radial compressibility as well as the normal degree of axial compressibility [95]. Finer denier fibers, however, will appear lighter in shade and require more dye to be matched to the same shade as with a coarser denier fiber [96]. More recently, some manufacturers that produce fabrics with silk-like or peach-skin-like appearances have introduced self-extensible filament yarns with apparently improved dyeing properties [30,97].

5.1. Problems originating from texturing filament yarns

Texturing is a manufacturing process that introduces bulk to an otherwise flat yarn. Continuous filament yarns are textured by several different methods such as false-twist, stuffer-box, knit-de-knit, gear crimping, edge crimping, and air-jet [98]. These techniques impart permanent helical deformation to the yarn. The filaments are crimped or looped at random to give the yarn greater volume or bulk; then, the yarns are heat-set by passing through a heating zone to generate dimensional stability. The most common form of texturing is false-twist, in which stacked friction disks or cross-belt assemblies are employed [21].

During the texturing process, the yarn is drawn further to impart additional strength and chain orientation to yarn before knitting or weaving processes. With an increase in the temperature of texturing chamber up to a certain point, dye uptake decreases, but beyond this point dye uptake begins to increase [99]. This increase in dye uptake results from crystal formation. An increase in crystallinity limits the available area where dye molecules can diffuse into the fiber. The increase in dyeability results from the melting and re-crystallization of numerous smaller crystals into fewer larger ones. When these small crystals melt and re-crystallize, the overall area for dye penetration increases. In addition, in friction disc texturing, dye uptake increases due to increase in disc spacing, the angle of yarns orientation on the disc, and from the increase in the surface helix angle of the twisted yarn. These changes can give rise to an increase in the twist inserted into the yarn that decreases crystallinity [100]. In order to obtain uniform dyeing on textured polyester materials, additional factors should also be taken into consideration, which include pretreatment, selection of dyestuffs and auxiliaries, dyeing cycle, and dyeing machinery [33].

5.2. *Package dyeing problems originating in yarn winding*

Package dyeing of polyester yarn provides the textile producer with increased flexibility and diversity in areas of fabric formation. Package dyeing is the process of batch dyeing yarn packages wound on perforated tubes where radial circulation of the liquor takes place. The principal mechanical factors influencing the rate of dyeing include [101–103] the following:

(1) Rate of liquor circulation.
(2) Rate of temperature rise.
(3) Porosity of package.
(4) Liquor viscosity.
(5) Density of package.

Flow rate and cycle time vary with fiber blends and dyeing machine used and trial and error is often employed to determine the best conditions for package dyeing.

The success of yarn package dyeing is dependent on the degree of the preparation of yarn packages. In some manufacturing processes, considerable economics can be made if the yarn wound on all packages is the same length and dyed uniformly. This is greatly influenced by the yarn winding process. The most important parameters affecting the winding process are as follows [101–103]:

(1) Winding type.
(2) Winding system.
(3) Winding ratio.
(4) Winding angle.
(5) Package density.
(6) Package shape.
(7) Cone volume.
(8) Spindle movement speed.

However, fiber hairiness can also influence the efficiency of the spinning and twisting processes. Increased hairiness has been found to be a main cause of dye strike in dyed packages. High degrees of hairiness on fiber surface may necessitate dye stripping after dyeing process. Fiber hairiness can be caused by the following [102]:

(1) Impact of the yarn balloon against the edge of overlapping divider plates causing periodic disturbance of yarn.
(2) Increasing the spindle speed on ring frames.
(3) Using different weights of ring travellers.
(4) Using different types of caps, e.g. straight and bell caps.
(5) Tacky or otherwise unsatisfactory roller coverings causing partial licking of the yarn.

Proper preparation of wound packages is very important for successful dyeing. Each package should be of the same size, density, and weight; otherwise the dye liquor follows the path of least resistance leading to unlevelness. Yarn packages should also be wound as softly as possible without sloughing to allow yarn to shrink without impeding the flow of dye liquor or deforming the package structure. In package dyeing, since yarn packages swell and shrink, a cushion on the dye tube will assist in obtaining level dyeing.

A well-wound package not only prevents hairiness and promotes level dyeing but also minimizes the risk of encountering several dyeing problems. Faulty packages with varying and improper density cannot be easily dyed to level shades even after taking all necessary precautions during dyeing. This is on account of large differences in the rate of flow of liquor caused by density variations. Too hard a package leads to uneven dyeing because of channelling of dye liquor; while too soft a package makes unwinding after dyeing difficult and may necessitate slowing down of machine speed and also increases the amount of hard waste produced [104,105]. In addition, uniformity of package density, both within the package and between packages, is critical to obtain uniform results. An obvious result of density variations between packages is shade variation from package to package and the nonuniform appearance of fabric. Several factors affect uniformity of wound package density, including winding tension, yarn count, yarn fineness, yarn twist, pressure between package and winding drum, type of supply packages, increase in package diameter, and yarn reversals on cross-wound packages [106].

Winding tension is determined by tension weights used on tensioning devices as well as the winding speed. Winding tension determines the tightness with which yarn is laid onto the package and thus package density is generally increased with an increase in winding tension. With conditions such as winding speed, tension weights, and package size remaining constant, *yarn count* determines the winding tension and spacing of yarn layers on the package. Both tension and spacing of yarn layers are reduced as the yarn becomes finer. Reducing tension decreases the package density, while reduced spacing of yarn layers increases the package density. *The pressure between a package and the winding drum* is decided by dead weights on the cradle or spring loading. Increasing the pressure leads to more compact yarn package formations. The *type of packages* supplied has some bearing on the pattern of the liquor flow. Cylindrical shape packages have been shown to be ideal for a uniform liquor flow throughout the yarn package. Packages prepared from ring bobbins are characterized by local variations in density while those prepared from winding exhibit fairly uniform density throughout the package. As the package is built up, the space between yarn layers is increased, which leads to lowering density [16]. However, package density is reduced with increase in its *diameter*. It has been observed that the decrease in package density is higher in the initial stages of winding. Therefore, the variation in density within a package will be reduced if a bigger diameter sleeve is used and the final package diameter is increased by the same amount. In the winding process, yarn is traversed from edge to edge with yarn reversals at the edge of the cross-wound packages. Since there is a greater thread density at the reversal points compared to the body of the package, the edges become hard. Thus, the presence of hard edges on yarn packages is the characteristic feature of cross-wound packages. Wound package faults like ribboning, soft nose or base, straight wind, etc. are instrumental in giving local variations in dye pick-up primarily because the faulty portions differ in density from rest of the package [9,76,107].

The following range of precautionary recommendations is listed to help overcome some of the problems because of winding [108]:

(1) The winding tension should be lowered by reducing tension weights and winding speed.

(2) Excessive tension variations between spindles should be avoided to ensure uniformity of density between packages.

(3) The pressure between package and drum should be lowered by reducing dead weights on package cradle.

(4) The compensating system, which lifts up the package progressively in relation to the increase in package diameter, should be checked to ensure proper functioning.

(5) The lateral displacement motion, which prevents formation of hard edges, should also be checked to ensure proper functioning.

(6) Yarn package faults, including ribboning, soft nose or base, straight wind, etc., should be minimized as much as possible.

6. Problems originating in fabric formation

Generally fabric constructions are classified in the following three categories: (1) nonwoven, (2) woven, and (3) knitted [51,109]. Nonwoven fabrics are produced directly from fibers without intermediate fiber consolidation steps such as spinning into yarn or thread [110]. The resulting structure and physical properties of nonwoven fabrics differ from woven textiles, even when the basic fiber is the same, making them especially well-suited for a wide variety of applications that do not require uniform coloring [110].

Woven fabrics are classified according to their weave pattern or structure and the manner in which warp and weft yarns cross each other. The three fundamental weave patterns, of which others are variations, are the plain, twill, and satin. In *plain* weave, also known as calico, tabby, taffeta, or homespun weave, the weft yarn passes over alternate warp threads, requiring two harnesses only. The relatively simple construction is suited for the production of inexpensive fabrics and printed designs. Variations are produced by the use of groups of yarns, as in basket weave and monk's cloth, or by alternating fine and coarse yarns to make ribbed and corded fabrics, as in the warp-ribbed Bedford cord, piqué, and dimity and the weft-ribbed poplin, rep, and grosgrain. The second primary weave pattern, *twill*, shows a diagonal design made by causing weft threads to interlace two to four warp threads, moving a step to right or left on each pick and capable of variations, such as herringbone and corkscrew designs. Noted for their firm, close weave, twill fabrics include gabardine, serge, drill, and denim. *Satin* weave pattern has floating or overshot warp threads on the surface that reflect light, giving a characteristic lustre. When the uncrossed threads are in the weft, the weave is called sateen [60,111,112]. The main factors that can give rise to uneven dyeing during the manufacture of woven fabrics are as follows [18,60,113–116]:

- Variation in stretch in the warp, or longitudinal, yarns.
- Variation in weft and warp yarn characteristics (e.g. twist, twist direction, count, hairiness, blend ratio composition, color, etc.).
- Variations in the weft, or filling yarn crossing the warp, binding the warp threads.
- Variation in picking, or inserting the weft; and battening.
- Variations in fabric compactness.
- Variation of frames holding the warp and throwing the weft yarn.
- Variation in size application.
- Fly or foreign fiber woven into the fabric.

Knitting is a fabrication process in which needles are used to form a series of interlocking loops from one or more yarns or from a set of yarns. Also, knitting is the most common method of interloping and is second only to weaving as a method of manufacturing textile products [117]. The main factors that can give rise to unlevel dyeing during the manufacture of knitted fabrics are [18,60,113–116]:

- Variation in fiber diameter, yarn linear density, and tightness factor.
- Variation in yarn characteristics (i.e. twist, twist direction, count, hairiness, blend ratio composition, color, etc.).
- Variation in the angle of spirality of yarn.
- Fly or foreign fiber knitted into the fabric.
- Variation in knitting machinery employed (manual or automatic).
- Variation in course length.
- Variation in wale density.

Environmental and storage conditions have an important effect on level dyeing of knitted goods. Unfavorable storage conditions such as high humidity and temperature could result in the formation of small dark areas, resembling water spots on dyed goods. In a reported case, clean prescoured knitted goods made of texturized polyester yarns, exhibited water spot like faults where microscopic examination of fault provided no hints for the cause of unlevelness [117]. Although the knitted material was clean and showed no encrustations or deposits, it was found that the dried fabric was stored under a shed roof where raindrops occasionally fell onto the fabric through an open window. The fabric, which became wet in some areas, was subsequently heat set in a stenter frame. After dyeing the heat-set fabric it was found that the stains only appeared in the wetted areas. Heat-set and partially wet fabric may have resulted in nonuniform setting, which can affect dye uptake.

In addition, fabric weight and structure exert a considerable influence on the rate of heating and level dyeing [95]. As the thickness of the fabric layer increases, a sharp increase in the heating-up time is observed. The increase in heating-up time because of an increase in the material weight, however, is not exceptional and is due to heat transfer principles.

In addition to the issues identified above, fabric yellowing is also often encountered as a common problem before the dyeing and finishing steps. Yellowing may be due to various reasons, including those listed below [3,20,108]:

- Air oxidation.
- Drying at high pH.
- Exposure to various gases, NO_2, and ozone.
- Use of acid sensitive brightener.
- Excessive use of brightener.
- Micro-organisms, algae, and humic acid.

7. Problems originating from pretreatment

As a term, preparation has two implications in textile processing. In greige manufacturing, this term is used to describe the process that is used to prepare yarns for weaving and knitting such as slashing of warp yarns prior to weaving. In dyeing and finishing, the term is used to describe those processes that prepare fabrics for the subsequent steps, as well as coloration and finishing. Fabric preparation is the first within a series of wet processing steps where greige fabric is converted into finished fabric. The steps that follow preparation, including dyeing, printing, and finishing, are greatly influenced by how well the fabric is prepared. Improper preparation is often the cause of problems encountered in the dyeing and finishing steps. Successful preparation depends on several factors, some of which are listed below [9,79,102,109,116]:

(1) Level and type of impurities present.
(2) Chemicals used in various stages of preparation.
(3) Water supply.
(4) Type of machinery used.

7.1. *Problems due to singeing*

Singeing is a process applied to both yarns and fabrics to produce an even surface by burning off projecting fibers, yarn ends, and fuzz. This is accomplished by passing the fiber or yarn over a gas flame or heated copper plates at a speed sufficient to burn away the protruding material without scorching or burning the yarn or fabric [20]. Fibers protruding from yarn and/or fabric appear darker after dyeing, leading to an inferior fabric appearance. Weaving defects, such as raised and depressed areas in the textile fabric, can cause difficulties during singeing and therefore should be minimized [7]. Singeing is usually followed by passing the treated material over a wet surface to ensure that any smouldering is halted. The singeing process improves several fabric characteristics, which includes increased wettability, improved dyeing characteristics, increased reflection because of minimizing the 'frosty' appearance and generating a smoother surface, better clarity in printing, improved visibility of the fabric structure, less pilling, and decreased contamination through removal of fluff and lint [9]. If singeing is to be carried out after dyeing, care must be exercised in selecting dyes to ensure they have good sublimation fastness. Although proper singeing is required to produce a uniform fabric surface, the sequence in which it is conducted is important. Various parameters may have to be taken into account for each type of fabric processed. These parameters include the following [29,118,119]:

(1) Flame intensity.
(2) Singeing speed.
(3) Distance between the flame and substrate.
(4) Method of singeing.
(5) Singeing position.
(6) Yarn package shape and type.

7.2. *Problems due to desizing*

Sizing agents are applied to reduce the frictional properties of warp yarns by coating their surfaces with film-forming polymers. The desizing procedure depends on the type of size. It is therefore necessary to know what type of size is on the fabric before desizing. Size removal depends essentially on the following factors [119–123]:

- Viscosity of the size in solution.
- Ease of dissolution of the size film on the fiber.
- Amount of size applied.
- Nature and amount of the plasticizers.
- Fabric construction.
- Method and nature of washing off.
- Temperature of washing off.

Starch is the most difficult size to remove. This is more important when dealing with blends of polyester and cellulose. Starch does not dissolve in water and must be broken down chemically into water-soluble compounds by enzymes, oxidizing agents, or acids. Typical desizing processes involve the following steps:
(1) Saturating the fabric with the following solutions at 160°F.

0.1–1.0%	Bacterial amylase enzyme
10.0%	Sodium chloride
0.5%	Calcium chloride
0.1–0.2%	Wetting agent (non-ionic)

(2) Digestion of the starch is carried out according to one of the following processes:

- Batching for 2–12 h.
- J-box dwelling for 10–30 min at 170–180°F.
- Steaming for 30–120 sec at 210–212°F.
- Thorough washing at 180°F or above.

The washing step is very important in removing the starch from the fabric [121]. After desizing, the fabric should be analyzed to determine the uniformity and thoroughness of the treatment. An alternative desizing procedure is oxidative desizing with peroxides. The oxidizing agent is hydrogen peroxide or in some cases a combination of hydrogen peroxide and potassium dipersulphate, and this process is done under alkaline conditions. Traditional desizing is performed by using acid or oxidative desizing agents that are associated with many drawbacks and limitations. However, because of its relatively uncontrolled and nonspecific reaction, the cellulose material also gets damaged and looses strength. With the introduction of enzyme-based desizing process, the limitations and drawbacks of traditional desizing process can be overcome. The enzymatic desizing process is performed by using α-amylase enzyme. The advantages of enzymatic desizing over traditional desizing include [119,121–123]:

(1) Better strength retention due to very specific reaction of enzyme with the sizing agent, which does not adversely affect cellulose.
(2) Saving of water as multiple washing is not required to remove the residual chemicals after desizing.
(3) Processing time necessary for desizing can be reduced.
(4) Neutralization is not required because the same processing conditions are required in the next process leading to zero salt formation in Effluent Treatment Plant (ETP).
(5) Savings in energy as desizing takes place at moderate temperatures.
(6) Softer fabric feel and less hairiness on the fabric.
(7) No adverse effect on other bath auxiliaries because of mild process conditions used.
(8) No adverse effect on fabrics containing Lycra or polyester.
(9) Safe and easy handling of reagent.
(10) No adverse effect on machinery.

There are several problems associated with the desizing process. Some of these issues and precautionary recommendations are listed below [119,121–123]:

(1) Incomplete and/or uneven desizing and washing off can adversely affect the subsequent caustic scouring process in which the fabric may shrivel to one side because of slipperiness compounded by the caustic bath.

(2) The size recipe applied on the warp, including information on the preservative used, softeners, other additives, etc. should be made available to the processor so that action could be taken to provide suitable desizing process and recipe.

(3) It is necessary to ensure that the sizing recipe does not contain any ingredients that cannot be made soluble and/or emulsified and washed off.

7.3. *Problems due to scouring*

The scouring process removes water insoluble materials such as oils, fats, and waxes from textile fibers. These impurities coat fibers and inhibit rapid wetting, adsorption, and absorption of dye and chemical solutions. In addition, the scouring process softens and swells motes, which facilitate their destruction during bleaching. Oils and fats are removed by saponification with a hot sodium hydroxide solution. Unsaponifiable materials such as waxes and dirt are removed by emulsification. Both of these processes (saponification and emulsification) take place within a typical scouring process. The typical scouring process involves the following steps [119]:

(1) Saturating the fabric with the following solution at 180°F:
  ○ 0.1–0.2% surfactant;
  ○ 0.3–0.5% sequestering agent;
  ○ 4.0% sodium hydroxide.
(2) Steaming using one of the following methods:
  ○ 60–90 min at 212°F in J-box, or
  ○ 20–30 min at 212°F in open-width box steamer.
(3) Thorough washing at 180°F.

Synthetic fibers such as polyester are scoured with milder formulations such as soap or detergent containing comparatively small amounts of alkali (0.1–0.2% sodium carbonate). The choice of auxiliaries such as surfactant and chelating agent is essential to obtain a complete scour. After scouring, the fabric should be checked for thoroughness of scouring. An alternative to aqueous scouring is solvent scouring. This process uses organic solvents such as perchloroethylene and trichloroethylene as a scouring medium instead of water. The oils, fats, and waxes on fibers are easily dissolved in these solvents. This process does not produce any fiber damage, and has low energy requirement. However, in textile processing mills most solvents are often not employed due to their volatility, health, and safety, as well as cost considerations. The important parameters that should be considered in the successful execution of the scouring process are as follows [119,123]:

• Concentration of alkali.
• Type of alkali.
• Temperature.
• Time.

The high concentration of alkali and high temperature in the scouring solution may lead to harsh surface of substrate and unlevel dyeing. The complete removal of the scouring agent from the substrate by wash-off stage should be considered to prevent dyeing defects.

7.4. Problems due to bleaching

The bleaching process destroys color impurities in fibers producing a white substrate. This is usually accompanied by treating the fibers with an oxidizing agent that oxidizes the color impurities to colorless compounds. Hydrogen peroxide, sodium hypochlorite, and sodium chlorite are the most common bleaching agents. When dealing with polyester and cellulose blends, fabric should be visually observed after bleaching, for presence of motes and pinholes. It should be practically free of motes. Inadequate mote removal usually results when the pH is too low during bleaching, peroxide decomposition during bleaching is high, or the motes are not adequately softened in scouring. Pinholes are the result of localized fiber degradation by hydrogen peroxide and are usually catalyzed by metal ions such as iron and copper.

A problem that is often detected after dyeing is the silicate deposits especially after dyeing to light shades. These deposits are seen as light patches or spots during bleaching. Silicate deposition can be detected by the formation of an intense yellow color in a molybdic acid (H_2MoO_4) test conducted on the ashed fabric. Deposition of silicates on the fabric can be caused by several factors that are listed below [124,125]:

- Improper ratio of Na_2O to SiO_2 in the bleach solution. The ratio should be approximately 1:1.
- Low pH during bleaching, since solubility of sodium silicate increases with pH in the alkaline range.
- Inadequate rinsing of bleached fabric.

Sodium silicate is difficult to completely rinse from fabric because the pH decrease with rinsing leads to decreased solubility of the silicate. Potassium silicate is more soluble than sodium silicate. Short fabric contact time provides more flexibility in choosing temperature and bleach and stabilizer concentrations, and minimizes or prevents fiber damage because of metal impurities and specific alkaline hydrolysis of polyester fibers [126]. Several types of synthetic fibers, including polyester, are bleached with peracetic acid (CH_3CHOOH) rather than with hydrogen peroxide because satisfactory reflectance values of fabrics may be obtained at a lower pH without causing fiber damage. The most common problems in bleaching with hydrogen peroxide are as follows [119,123,125]:

- Inadequate formation of latent deposit.
- Low degree of whiteness.
- Uneven whiteness.
- Broken or torn-off substrate.
- Reduced mechanical strength.
- Catalytic damage.
- Resist mark formation.

The disadvantages in bleaching with sodium chlorite are as follows:

- A possibility of liberating the toxic gas, chlorine dioxide.
- The equipment is expensive because of the need for exotic construction materials.
- No rapid bleach process available.
- Poor absorbency because of residual fat and waxes.
- Incompatible with most dyes and fluorescent brightening agents (FBAs)
- Use of multi-chemical baths, which need control.

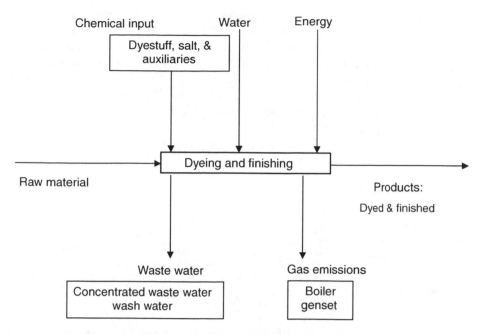

Figure 10.   Schematic diagram of water usage in dyeing and finishing processes.

Thus, fibers subject to alkaline hydrolysis and degradation may be bleached with peracetic acid at pH of 6.0–8.0 without appreciable fiber degradation or hydrolysis. The mechanism by which peracetic acid bleaches fibers is thought to be similar to that of hydrogen peroxide, since both are catalytically decomposed by metal ions and stabilized by sequestering agents such as polycarboxylic acids.

8. Problems originating from water quality

Water is the most important processing medium in a dye house. Definitely, the highest water use in textile wet processing is for the general purpose of washing and in preparation (desizing, scouring, bleaching), as well as dyeing as shown in Figure 10. Water is universally used as a solvent for the coloration of polyester and has a strong plasticizing action on the fiber. The quality of water can be varied depending on the treatment methods used. The following approaches have been recommended [90,107,127–129]:

- Water should be filtered using sand filters to remove the suspended solids.
- It should be neutralized to a pH of around 7.
- It should be treated, if necessary, to remove residual chlorine.
- It should be softened to remove and precipitate heavy metal ions.

Various textile wet processes are influenced in different ways by the presence of impurities in the water supply. The presence of chlorine, iron, and treatment chemicals commonly found in water can have a major effect on dyeing process. Metals can be introduced from piping system, the boiler through live steam, or dyeing equipment. Water for dyeing is usually chlorinated; it can change the shade of dyed fabrics or cause other effects on processing. This often contributes to poor color yield and laboratory to dye house correlations for dye recipe. Some of the most frequently observed dyeing defects can arise from water contamination [130]. These include inconsistent shade, blotchy and/or streaky dyeing, filtering, spots, and resist marks. Shade variation can be caused by chlorine contamination of water. This can cause color loss for many classes of dyestuff. Shade variation can also be caused by metals such as iron, copper, and other metals. These metals are known to affect many dyes, especially disperse dyes. Traces of metals can change the shade of disperse colors. For example, the colors of azo dyes are reduced by metallic salts and anthraquinone dyes are chelated with metals. The most common metals affecting shade variation of textile goods in water are aluminium, copper, manganese, and iron [90,110,127–129].

The most common impurities present in water that contribute to water hardness and affect shade variation are as follows [90,110,127–129]:

- Alkalinity and acidity.
- Suspended matters (oil, grease, clay, and sediments).
- Calcium and magnesium.
- Residual soaping/rinsing agents.
- Silicate and silica.
- Sulphates.
- Chloride.
- Miscellaneous anions (sulphide, fluoride, etc.).

Table 1 shows a list of disperse dyes and their susceptibility to shade change in the presence of metals such as iron and copper. The presence of calcium and magnesium in the process water can cause uneven and unbalanced washing off of unfixed dyes. This can lead to streaks, blotches, and inconsistent shade. Water conditioners such as hexametaphosphate (hexaphos) are effective and are generally safe in the sense that they do not cause other undesirable effects such as shade variations. Blotchy or streaky dyeing can result from acidity and alkalinity in the water, which can affect exhaustion, levelling, and fixation of dyes. These problems can be avoided by several ways, including the judicious use of chelates and/or dispersants, dye selection, or water purification prior to use.

Depending on sources of water used, it is critical to know whether chlorine is present in the dyebath employed for the dyeing of polyester fibers with disperse dyes. Sodium thiosulphate is the preferred antichlor agent used to remove traces of chlorine present and

Table 1.   Disperse dyes affected by metals.

Anthraquinone dyes that chelate with metals	Disperse dyes that do not chelate with metals	
C.I. Disperse Red 3	C.I. Disperse Red 65	C.I. Disperse Red 338
C.I. Disperse Red 55	C.I. Disperse Red 73	C.I. Disperse Red 339
C.I. Disperse Red 60	C.I. Disperse Red 88	C.I. Disperse Green 9
C.I. Disperse Red 91	C.I. Disperse Red 90	C.I. Disperse Blue 60
C.I. Disperse Red 116	C.I. Disperse Red 117	C.I. Disperse Blue 148
C.I. Disperse Red 263	C.I. Disperse Red 153	C.I. Disperse Blue 165
C.I. Disperse Blue 27	C.I. Disperse Red 177	C.I. Disperse Blue 183
C.I. Disperse Blue 56	C.I. Disperse Red 274	C.I. Disperse Blue 291
C.I. Disperse Blue 73	C.I. Disperse Red 305	C.I. Disperse Blue 337
C.I. Disperse Blue 118	C.I. Disperse Red 307	

Note: Yellows are generally insensitive to metals.

its slight overuse does not harm the dyes. Water hardness and the presence of metals can also affect some dyes. When water hardness is suspected, the use of sequestering agent such as ethylenediaminetetraacetic acid (EDTA) is recommended.

Sequestering (chelating) agents can be defined by their ability to form a complex with metal ions, which allow these ions to remain in solution despite the presence of precipitating agents. They are normally applied to mask hardening alkaline-earth cations and transition metal ions in aqueous solutions, in order to eliminate their damaging effects, which include the following [110,119,131]:

(1) Catalytic destruction of hydrogen peroxide.
(2) Potential inactivation of the enzymes in enzymatic desizing, resulting in poor size removal.
(3) Precipitating soaps in scouring processes (especially calcium and magnesium ions), forming a sticky insoluble substance. The sticky deposits impair fabric handle, cause resist spots in dyeing, attract the degree of soiling of the material, and cause inconsistent absorbency in subsequent processes.
(4) Drastically reducing the solubility and the rate of dissolution of surfactants due to formation of complexes with alkaline and alkaline earth salts; thus impairing the wash removal ability of the surfactants.

When metal-containing dyes are in use, the application of EDTA should be avoided because it can strip the metal form the dye molecule. Sodium hexametaphosphate can be used to control water hardness where EDTA cannot be used [77]. Reduction decomposition can be caused by ionization of metal or stabilization of reduction-causing metal ions. Some dyes are decomposed by the reducing power generated by the ionization of metals. Some metal ions deteriorate the stability of dye dispersions leading to low color yields. Formation of a complex between metal ions and disperse dyes depends on the chemical structure of dyes. Especially, the shade of red anthraquinone dyes changes markedly to bluish as a result of complex formation, which causes shading issues in dyeing. The shade change because of the formation of complexes depends on the metal: Fe^{2+}, Fe^{3+}, and Cu^{2+} complexes change markedly, but the change because of complex formations with Mg^{2+} and Cu^{2+} is to a smaller extent. In order to prevent the influence of metal ions, the addition of a sequestering agent is effective. For Fe^{2+}, Fe^{3+}, and Cu^{2+}, EDTA and nitrilo triacetic acid (NTA) are used, and for Mg^{2+} and Cu^{2+}, polycarbonic acid-type are mainly used.

The influence of reduction caused by the ionization of metals in a dyebath cannot be avoided by sequestering agents, and should be considered separately. For example, if iron powder is present in the dyebath, Fe^{2+} or Fe^{3+} may be generated. A sequestering agent can catch such ions after ionization. However, even after the formation of a complex between Fe^{2+} and the sequestering agent, the product still has reducing power. Therefore, reducing power cannot be neutralized by the sequestering agent only. In order to prevent the adverse influence of metals' reducing power, the addition of an oxidizing agent is recommended, but the optimum amount of this agent should be decided carefully. For example, for a dyebath including 50 ppm of Fe^{2+}, the addition of 1 g/L of sodium chlorate is effective. However, an excess amount of chlorate leads to the decomposition of the dye by oxidation; thus, the use of chlorate is not generally recommended. If the amount of metal in the dyebath is small, anti-reducing agents such as sodium meta-nitrobenzene sulphonate could be employed. Decomposition by reduction of the dye by iron powder or Fe^{2+} is subject to occurrence of acidic conditions. Beside metals, some auxiliaries with reducing power often cause dyeing troubles. Therefore, if susceptible dyes to reduction are used, auxiliaries with nonreducing power should be selected carefully.

9. Problems originating from heat setting

Heat setting may take place before or after dyeing and there are advantages and disadvantages to both. Heat setting prior to dyeing is desirable and indeed considered by some to be essential to impart stability to polyester fibers, especially if the temperature used during dyeing exceeds the normal atmospheric boil. Heat setting polyester fabrics is commonplace in today's industry because this process imparts dimensional stability and improves crease and wrinkle resistance. The following are the advantages of heat setting before dyeing [9,94,132]:

(1) Dyeing can be carried out in rope form at high temperature under pressure without fabric deformation.
(2) A variety of disperse dyes can be used since they will not be subjected to high heat setting temperatures.
(3) Dyeing following heat setting can help remove the stiff hand imparted by heat setting.

Following are the disadvantages of heat setting before dyeing [9,94,132]:

(1) Any impurities not removed during the scouring process may be permanently set into the fabric creating dye resist or dark spots.
(2) Any unlevel temperature during heat setting will cause variations in the shade of the dyed fabric.
(3) If fabrics sized with polyvinyl alcohol are heat set before desizing, some of the size may be permanently set into the fabric.

The advantages of heat setting after dyeing are as follows:

(1) Unlevel dyeing that results from nonuniform heat setting temperatures can be avoided.
(2) Energy savings result because the fabric does not need drying after preparation.

Disadvantages of heat setting after dyeing are as follows:

(1) Creases can be created during dyeing, which cause unlevel dyeing.
(2) Only certain disperse dyes can be used, which will not volatilize during heat setting leading to shade differences.
(3) Thermo-migration of dyes may take place, which causes the dye to no longer remain fixed inside the fiber. This problem will drastically affect the fastness properties of the fabric.

Heat setting alters the internal molecular structure of the fiber causing the amorphous regions to become more oriented. This increase in orientation limits the number of polymer chains that can move. It is this chain movement that allows dye molecules to enter the fiber [9]. The heat setting temperature is usually 130°C; which is above the temperature the fabric will experience during further processes [56]. Above 165°C, dye uptake increases because crystalline areas become less oriented. This decrease in crystal orientation allows more polymer chain movement; therefore, an increased area is exposed for dye molecules to penetrate the fiber.

10. Problems originating from steaming

Steaming is one of the most important processes in textile finishing, and is widely used in preparation and dyeing. Controlled temperature, moisture, and amount of air in machine are the main parameters that should be controlled to get an ideal steaming process. Problems originating at the steamer include water spots and loss of depth or change of shade. The steam should be saturated and shall not dry the fabric. Superheated steam can cause defects in the fabric. Even the best steamer cannot produce high-grade fabric if the steam used is superheated. In the alkaline scouring step, even slight superheating of the steam can cause migration of the caustic soda in fabrics, leading to permanent marks and uneven dye uptake in dyeing processes [133,134]. Superheating of the steam also results in instability of dimension and lint. This may be detrimental to smooth, crease-free fabric running. Therefore, the effectiveness of the steaming process depends largely on achieving and maintaining the supply of steam at a constant rate and air detection at some point [135,136]. Modern steamers come with instrumentation to control the levels of chemicals and water. The use of a specially designed inlet to avoid ingress of cool air and shorter distance between the guide rollers to avoid crease formation are the other parameters that ought to be considered.

11. Problems originating from other wet processes

Controlling the dyeing process itself is of prime importance and critical factors influencing the dyeing process must be identified. These usually include pH, time, temperature, liquor ratio, electrolyte, and auxiliaries. Adding excessive chemicals to the recipe will increase cost and may cause quality problems. All parameters, including the temperature and amount of water added to a bath, should be documented and provided to mill operators responsible for mixing the recipe.

The use of polyester fibers in many textile applications is growing very rapidly because of their high strength, good elastic recovery, and dimensional stability after heat setting as well as suitability for blending with natural fibers [103]. However, the main drawbacks in polyester-based textiles, which include, for example, low moisture content, static accumulation, soiling, uncomfortable feel, pilling tendency, and difficulty in dyeing, are attributed to their high crystallinity, compactness, hydrophobic nature, and absence of chemically reactive groups [137]. Therefore, considerable efforts and technical developments have been extended to upgrade their quality and usefulness, and these include carrier-free dyeing, package dyeing, incorporation of anionic or cationic active sites in the fiber, Thermosol™ dyeing, and dyeing in an alkaline medium [138].

11.1. Problems caused by foaming

A yarn package contains a significant amount of entrapped air. When introduced into a dyebath containing a surfactant, the air produces a finely dispersed foam cushion as soon as dye liquor begins to flow through the package. Foam prevents the flow of liquor through the yarn and the resistance to flow may become so great as to deform or even burst the package [31,139]. In high-temperature jet dyeing machines there is considerable turbulence and foam is produced preventing dyebath–fiber contact and causing uneven dyeing results. Also, in padding the dry fabric passes rapidly through a dyebath containing the dye or chemical to be applied. If the bath contains a surfactant, air introduced by the fabric produces foam, which prevents intimate contact between solution and fiber even during squeezing between the padding fibers. In order to prevent or reduce the formation of foam during or after textile processes, the use of anti- and de-foaming agents is recommended.

As these agents are almost insoluble in water, they are applied from emulsion. When such emulsion is very stable, the agent is not released for adsorption at the foam interfaces and it cannot fulfil its purpose [117,140].

Foam may also be applied to textile substrates by means of an air knife, with the knife forming a 30–45° angle to normal, in which ram pressure can be easily built up. The air knife not only applies the foam superficially but also presses it deep into the textile good. The generation of foam in textile processing because of the presence of surfactants causes a number of problems. The need to control foam during dyeing and finishing is well recognized. Excessive foam can cause [141]

- floating and tangling of goods,
- uneven dyeing, and
- nonuniform application of finishes.

Anti-foams are commonly used in jets to decrease foam resulting from turbulence inside the jet. Even though anti-foams are included in most of the formulas, it is best to find the foam producer and remove or change it rather than introduce another dyebath auxiliary [142].

11.2. Problems caused by dyeing processes

The details of the dyeing process can vary considerably between different types of textile materials employing different types of dyeing equipment. For example, the maximum permissible rate at which the temperature of the bath can be raised may be determined by the relationship between the rate of circulation of the dye liquor and the rate of transfer of the dye from bath to fiber such as in the dyeing of yarn in hank dyeing machines. Technical problems abound during dyeing of polyester and its blends. The preparation of stock solution for dyeing should be carefully handled to avoid creation of unpleasant odor, which can cause health concerns, pollute the air, or diminish water quality.

The criteria for choosing a suitable dyeing process vary and may include the following points [60,143]:

- Shade range.
- Fastness requirements.
- Quality requirements and control.
- Cost.
- Equipment availability.
- Selection of the dyes.

There are three main types of processes for the dyeing of polyester materials: (1) *batch*, (2) *continuous*, and (3) *semi-continuous*. Table 2 shows general dyeing processes according to the specific end products. *Batch* process is the most common method used to dye textile materials and often depends on the type of equipment available and the weights or lengths of the material to be dyed. Batch dyeing is often called exhaust dyeing because the dye is gradually transferred from a relatively large volume dyebath to the material being dyed over a long period of time. Generally, flexibility in color selection is better and cost of dyeing is lower for a textile product if the dye application is closer to the end of the manufacturing process. Polyester and some other synthetic fibers dye more easily at temperatures higher than 100°C. Batch dyeing can give rise to some bruises and pilings because of hydraulic and mechanical impact during the high-temperature dyeing process [116,135].

Table 2.   General methods used for the dyeing of polyester and its blends.

Dyeing methods		
Batch dyeing	Dipping method	High/normal-pressure jet dyeing
		Jig (or jigger) dyeing
		Beck (or winch) dyeing
	Padding method	Cold pad-batch dyeing
Continuous dyeing	Hot flue dyeing	
Semi-continuous dyeing	Pad-jig dyeing	

Batch dyeing involves applying a dyestuff from a solution or suspension at specific liquor to goods ratio. At the end of the dyeing operation the spent dyebath liquor is drained out of the vessel. The post-dyeing stage consists of washing with water to remove unfixed amounts of dyestuff from the textile substrate. Atmospheric becks can be used for dyeing at temperatures up to the boil ($\sim 100^{\circ}$C). Pressurized becks are used for dyeing at temperatures higher than 100°C. The greatest advantages of becks are their simplicity, versatility, and relatively low price.

Beck dyeing machines subject fabrics to relatively low lengthwise tension and encourage the development of yarn and fabric crimp. However, becks can cause abrasion, creasing, and distortion of some fabrics. Jet dyeing machines resemble becks in that a continuous loop of fabric is circulated through the machine. However, the cloth transport mechanism is dramatically different between machines. Jet dyeing machines provide the following advantages compared to atmospheric becks for dyeing fabrics made from texturized polyester [18,30,117]:

(1) Vigorous agitation of fabric and dye formulation in cloth tube increase the dyeing rate and uniformity.
(2) Rapid circulation of fabric through the machine minimizes creasing because the fabric is not held in any one configuration for very long.
(3) Lengthwise tension on the fabric is low, so the fabric develops bulk and fullness of handle.
(4) Dyeing at high temperature of about 130°C gives rapid dyeing, improves dye utilization, fastness properties, and makes possible the elimination of carriers required when dyeing at lower temperatures.
(5) The lower liquor ratio used in jet dyeing allows shorter dye cycles and saves chemicals and energy.

However, jet dyeing machines provide the following disadvantages compared to atmospheric becks for dyeing fabrics made from texturized polyester [18,30]:

(1) Capital and maintenance costs are higher.
(2) Limited accessibility makes cleaning between dyeing and sampling for color during the dye cycle difficult.
(3) The jet action tends to generate foam in dye formulations in partially flooded jet machines.
(4) The jet action may damage the surface of certain types of fabrics.

The term *package dyeing* usually refers to dyeing of yarn that has been wound on perforated cones so that dye liquor can be forced through the package. Packages may be tubes, cheeses, or cones. Liquor ratio in a package dyeing machine is typically about 10:1

when the machine is fully loaded. The use of lower liquor ratios can save water, energy, and chemicals. If the liquor covers the yarn package but does not fill the top dome of the machine, the liquor ratio would be slightly lower than it would be in fully flooded machines.

Continuous dyeing is suitable for woven fabrics. Most continuous dye ranges are designed for dyeing blends of polyester and cotton. Continuous dyeing is operated at constant composition of dyebath, in which a long length of textile fabric is pulled through each stage of the dyeing process. In general, these techniques operate at high dyestuff concentrations of 10–100 g/L but hardly generate any waste beside those due to equipment clean-up discharges. A continuous dye range is efficient and economical for dyeing long runs of a particular shade. Tolerances for color variation must be greater for continuous dyeing than for batch dyeing because of the speed of the process and the large number of process variables that can affect the dye application. The polyester fibers are dyed in the first stages of the range by a pad-dry-thermofix process. The cotton fibers are dyed in the latter stages of the range using a pad-steam process. Fabrics previously prepared for dyeing enter the dye range from rolls. Uniformity of dye application requires that continuous dyeing be done in open width. Typical line speed in a continuous dyeing process is 50–150 m/min. Padding is a critical step in continuous dyeing. The disperse dye formulation (and sometimes the dyes for the cellulosic component) is applied in the first padder. The fabric is immersed in the dye formulation usually at room temperature and squeezed to give a uniform add-on of dye formulation across the width and along the length of the fabric. Low temperature in the formulation in the padder minimizes tailing. Higher temperature promotes wetting of the fabric in the short time the fabric dwells in the pad formulation. The wet fabric leaving the padder contains dye uniformly deposited on the fabric. Radiant predrying using infrared energy inhibits migration of the dye. Drying is completed using steam-heated cylinders. A thermal treatment called thermosol fixes the disperse dye on polyester fibers. The thermosol oven heats the fabric to a temperature of 390–430°F, the exact temperature depending on the particular dye being applied. The fabric dwells in the thermosol oven for about 1–2 min. The dye sublimes and diffuses into polyester fibers during the thermosol treatment.

The function of the cooling cans is to lower the fabric temperature so that it does not heat the solution in the chemical pad. The chemical padder applies the dyes into the fibers. The steamer heats the wet fabric so that the dye can diffuse into the fibers. The fabric usually dwells in the steamer for 30–60 sec. The washing section of the range is used for rinses and chemical treatments, which may be required to complete the dyeing process, and washing of the fabric to remove unfixed dye and chemicals used in dyeing [68,90,124,127,128,139,144,145].

Semi-continuous dyeing is characterized by performing dyeing in a continuous mode but fixation and washing steps are run discontinuously. In this process the fabric passes through a padding machine where it is impregnated in the dyebath, and then the dyestuff is fixed on a jigger. Sometimes the fabric can be dried in a hot-flue drier after padding and before entering the jigger. The application of the dyestuff by padding allows for homogeneous dyeing and time saving when compared to traditional jigger dyeing. The advantages of semi-continuous dyeing methods for polyester are as follows:

(1) Suitability for smaller lengths dyed to a given color that would be insufficient to justify continuous processing.
(2) Optimum use of readily available equipment.
(3) High productivity and better uniformity at the padding stage than batch dyeing.
(4) Continuity of color over long runs compared with batch dyeing.

11.3. Problems caused by reduction-clearing process

The dyed polyester should be cleared of surface-deposited dye as well as auxiliaries by means of treatment with detergent or by reductive or oxidative treatments, in order to secure optimum fastness of dyeing. Customer complaints of low fastness to perspiration or water in polyester are always associated with the presence of loose dye particles on the fiber surface. Fastness problems arising from the presence of surface-deposited dye are most commonly associated with deep colors and has been observed that it is very difficult to preserve surface cleanliness in navy blues, bottle greens, and strong browns [42].

Owing to the use of insoluble, or sparsely soluble, disperse dyes on polyester, which tend to aggregate at 100°C or above and deposit on the fiber, and also due to the low rate of diffusion of dyestuff into the fiber from the fiber surface, the rubbing and wet fastness properties can be poor. Loose dye is generally encountered after dyeing heavy-depth shades, but may also occur with other shades when dyeing has been carried out at too low temperature, or for an insufficient length of time [69].

In order to enhance fastness properties and improve brilliance of shade of dyed polyester fibers, the reduction-clearing process is carried out. The reduction-clearing process makes use of the hydrophobic property of polyester fibers, which prevents the penetration of the fiber by almost all ionized water-soluble reagents at temperatures below the boil, and of the very slow migration from the fiber of disperse dyes at temperatures below 80°C [4]. A solution of caustic soda, sodium hydrosulfite, and Lissolamine RC (a cationic surfactant) has been found to be the most efficient method for this reduction-clearing process. The reduction-clearing process is carried out before the normal soaping treatment, which is recommended for all dyeings on polyester fibers. However, in the presence of caustic soda, the excessively severe reduction-clearing treatment can produce tendering of polyesters, and this tendency is increased by the presences of Lissolamine RC. In order to reduce any risk of damage to the fiber during the reduction-clearing process, the temperature of treatment recommended above should not be exceeded. Control of the reduction-clearing stage is influenced by the following four factors [26,81]:

(1) Temperature.
(2) Flow rate of incoming clean water.
(3) pH.
(4) Substantivity of hydrolyzed dye species.

In addition, the after-clearing operation should not adversely affect the desirable properties of the polyester fiber. Specifically, the process should not result in the following [146]:

- Corrosion of equipment: it should not be an irritant to represent any other danger to the health of the operators.
- Generation of hazardous substances: it should not generate hazardous substances in its reaction with unfixed dyes and additives.
- Fastness problems.
- Effluent-disposal problems.
- Unjustifiable economics.

12. Problems due to application of disperse dyes

The Society of Dyers and Colourists (UK) defines *disperse dyes* as substantially water-insoluble dyes having substantivity for one or more hydrophobic fibers, e.g. cellulose acetate or polyester, and usually applied from fine aqueous dispersion. The dyeing property of fibrous polymeric materials depends on the relative amount of the amorphous phase and the chain packing, especially in the intermediate phase between the crystalline and amorphous phases. Because of its rigid structure, well-developed crystallinity and lack of reactive dye sites, PET absorbs very little dye in conventional dye systems. Disperse dyes are the most important class of colorants for dyeing polyester fibers. Many commercial disperse dyes are produced in a metastable crystalline form, which gives higher dye uptake than the thermodynamically stable form [147]. Disperse dyestuffs are absorbed and retained through a process of solution in the less crystalline regions [17]. Their saturation solubility in polyester is high. A considerable amount of research has been extended to improve the dyeability of polyester fibers. Polymerization of a third monomer, such as dimethyl ester, has successfully produced a cationic dyeable polyester fiber into the macro-molecular chain. This third monomer introduces functional groups and dye sites to which the cationic dyes can be attached [148].

A dyeing process for polyester fiber at low temperature (40°C and below) has been reported [120]. This method employs a disperse dye in a microemulsion of a small proportion of alkyl halogen and phosphoglyceride. The main advantage of this method is low temperature processing, but there remains the environmental problem that is produced by using toxic carriers. The textile industry uses large amounts of water in dyeing processes emitting organic compounds into the environment. Owing to this problem a dyeing process for polyester fiber has been developed in which supercritical carbon dioxide (CO_2) is used as a transfer medium [149]. This method gives an option to eliminate water discharge, which is low in cost, nontoxic, nonflammable, and recyclable. When dyed in an aqueous medium, reduction clearing is to be carried out to stabilize color intensity, which produces more wastewater. However, the main disadvantage of supercritical CO_2 method is the high pressure used in the dyeing and cleaning vessel. However, equipments designed for work under elevated pressure are more expensive than those employed for conventional applications at atmospheric pressure [150].

In practice, the major cause of color mismatch is often misformulation [151]. A mismatch can occur as a result of the loss of one or more of the dyes from the bath by agglomeration or, more commonly, by decomposition [4]. Therefore, the selection of an appropriate disperse dye is very important for obtaining a level and well-penetrated polyester in the most economical manner available. Some of the faults in the dyeing of polyester with disperse dyes include the following [58,119,152–154]:

- Ending and tailing due to unstable color dispersion or combination of dyes of different migration properties.
- Ring dyeing mostly because of variation in stretching during texturing.
- Skitter dyeing because of inadequate preparation, quality variation of fiber, poor solidity of shade in blends, and irregular yarn tension in woven and/or knitted form.
- Localized pale dyeing because of air pockets in the materials.
- Precipitation of dye when the dyeing depth exceeds the saturation value of the fabric.
- Blue spots on red dyeings because of the use of higher dosage of anthraquinone disperse dyes or the presence of greasy impurities on goods.
- Low color build-up and low brightness because of alkaline water process.
- Poor light fastness in carrier dyeing because of residual phenolic-type carrier.
- Listing because of inconsistent heat setting and irregular tension in weaving of faulty batches.
- Center selvedge variation because of poor winding, irregular liquor flow, incompatible selection of dyes, or poor heat setting.

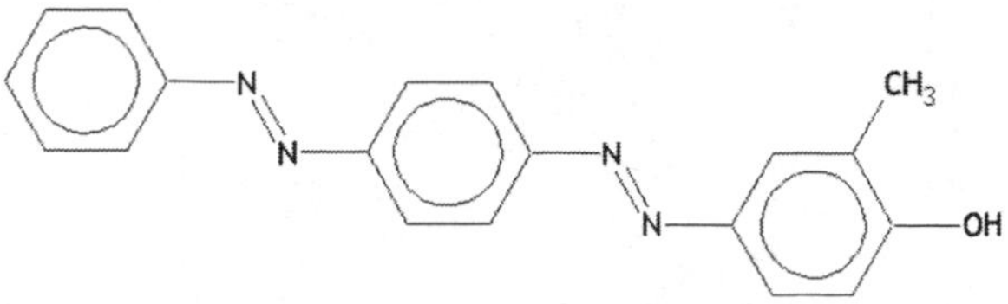

Figure 11. C.I. Disperse Yellow 7 (azo).

12.1. *Dye parameters*

12.1.1. *Disperse dyes*

This class of dyes has historical significance because it was the first class that was specifi-cally designed for a particular fiber. Disperse dyes were originally developed for secondary cellulose acetate because this fiber exhibited low water absorption, and alternative methods for improving the dyeing of acetate had proved unsuccessful [92]. By definition, *disperse dyes* are a class of substantially water insoluble dyes, originally introduced for dyeing cellulose acetates, and usually applied from a fine aqueous suspension [4]. The majority of disperse dyes belongs to three main groups based on their chemical constitution: (1) azo, (2) anthraquinone, and (3) nitrodiphenylamine. Azo dyes constitute the largest and the most important group of disperse dyes and provide mainly yellow, orange, red, and some violet and blue dyes and their water solubility is generally more than that of anthraquinones [155]. Anthraquinones yield bluish-red, violet, blue, and bluish-green dyes. Nitrodipheny-lamines mostly produce yellow and orange-yellow dyes [156]. Figure 11 represents a typical azo dye, Figure 12 shows an anthraquinone, and Figure 13 depicts a nitrodiphenylamine dye.

Today azo disperse dyes are becoming more prevalent because of expensive manufac-turing costs and environmental problems associated with anthraquinone dyes. However, level dyeings are achieved more easily with anthraquinones; therefore, light shades are almost always produced with this dye class, while azos are used for heavy shades whenever possible [4,92,110]. When using disperse dyes in jets, the dyes must possess the following criteria to yield level and reproducible dyeings [30,110,157]:

- Good dispersion stability at high temperatures with rapid agitation.
- Good levelling during the heating-up phase, and uniform dye uptake during the exhaustion phase.
- Easy reproducibility of shade with slight variation in temperature and rate-of-rise.
- Provide the shortest time cycle possible.

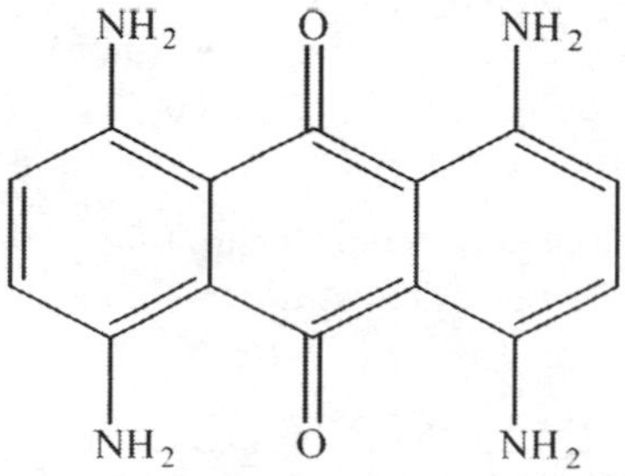

Figure 12. C.I. Disperse Blue 1 (anthraquinone).

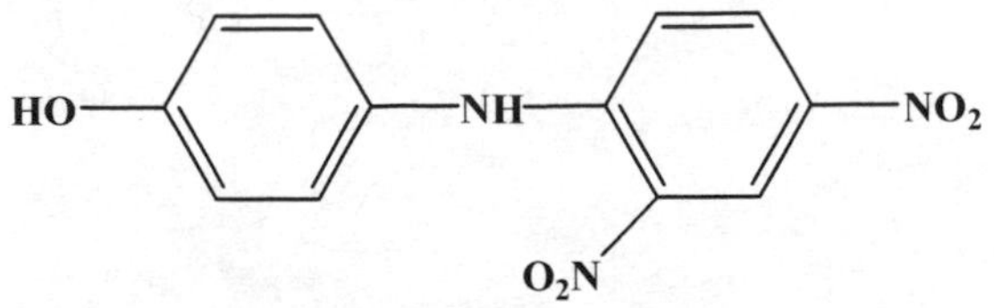

Figure 13. C.I. Disperse Yellow 1 (nitrodiphenylamine).

As has been noted, fiber blends have become increasingly important in recent years from commercial and technical standpoints. One of the most important blends in the apparel industry involves blending various ratios of cotton with polyester. If conventional azobenzene disperse dyes are used for the coloration of polyester/cotton blends, this can result in the staining of cotton to some extent, and the removal of this stain can be difficult. This problem can be minimized by the introduction of alkoxy carbonyl groups into the coupling component during the manufacture of dyes. A typical example of this type of dyes is shown in Figure 14. In the presence of alkali, the alkoxycarbonyl groups can be hydrolyzed to the corresponding carboxylic acids, which give the dye water solubility but with little or no substantivity for cotton or polyester fibers.

12.1.2. Aqueous phase transfer

Dispersion stability is the extreme case of imbalance between dye agglomeration and dye diffusion. The consequences of such instability are known as unlevelness and insufficient fastness [135]. Correcting these shortcomings is very expensive. At the end of a dyeing process, the dye that has been absorbed by the fiber is in a state of dynamic equilibrium with the dye that remains in the bath, and the fraction of the latter that is in aqueous solution must be present in the same state of aggregation as the dye in the fiber. The aqueous phase transfer of disperse dyes can be described in the following four stages [30,110,157]:

(1) Some of the dye dissolves in the water of the dyebath.
(2) Molecules of dye are transferred from solution to the surface of the fiber.
(3) The solution in the dyebath is replenished by the dissolution of more solid material from the dispersion.
(4) The adsorbed dye diffuses mono-molecularly into the fiber.

This mechanism mainly considers the sequence of events for a single fiber. In bulk form, the exhaust process may be considered to consist of three aqueous phase transfers, which can be listed as follows [30,110,157]:

(1) The transfer of dye from bath to fiber.
(2) The completion of dye diffusion (fiber penetration).
(3) The redistribution of dye between portions of the fiber that have unequal amounts of absorbed dye during the first phase.

The uniformity of the initial aqueous phase transfer is influenced by the relationship between the dyeing rate and the rate of redistribution of the dye liquor throughout the bath,

$$DYE-N-(C_2H_4COOAlk)_2$$

Figure 14. Typical structure of a new class of azobenzene disperse dyes.

by the pumping speed in the case of liquor circulating machines or by the speed of transport of cloth in other cases. For the dyeing to be level and even, it is essential that the color dispersion remains stable during the dyeing cycle. Some of the factors leading to instability of dye dispersions are [58,119,152–154]

- unfavorable pH (above 6);
- low liquor ratio;
- high dye concentration;
- adverse influence of auxiliaries such as electrolytes, dispersing agents, levelling agents, etc.;
- use of fibers of low absorption capacity;
- liquor turbulence as in jet dyeing; and
- use of red, purple, and blue disperse dyes of anthraquinone structure, which are susceptible to metal ions present in water (where sequestering agents should be added to the dyebath).

12.1.3. Dye-fiber substantivity

The affinity of a dye for a given substrate through interactions such as polar or ionic attractions, hydrogen bonding, and Van der Waals forces is termed as its *substantivity*. High dye substantivity means that the fibers become fully penetrated by the dyes within a few seconds of having reached the operating temperature presumably because both fiber and dye have similar mean energy densities. Some authors have previously proposed that the adsorption of disperse dyes on hydrophobic fibers can be accounted for by the combination of strong hydrogen bonding and weak dispersion forces [110,154]. Also, hydrogen bonding has been mentioned as a more important contributor toward dye–fiber interaction than dispersion forces. In other words, dipole–dipole and dipole-induced dipole forces of interaction may operate between the fiber and the dye [57]. Therefore, the substantivity of disperse dyes toward polyester fiber can be attributed to a variety of interaction forces, which will in turn affect various dyeing properties. In the case of dyeing of cellulosic fibers with anionic dyes, generally, the pH level of the dyebath solution in dyeing process is very high, which results in an electrostatic repulsion force operating between anionic dye and fiber. In the case of polyester/cotton blends, by adding an appropriate amount of salt such as sodium sulphate or sodium chloride to the dyebath, the electrostatic barrier between the fiber surface, for cotton component, and the dye, known as the *Donnan potential*, can be suppressed, which helps to facilitate dye/fiber contact, hence, improving substantivity. At equilibrium, the counter-ion sodium is more concentrated on the fiber than the anion sulphate or chloride [4,110]. Care has to be exercised, however, to prevent the formation of ring dyeing when the substantivity of the dye to the fiber is high [158]. In dyeing polyester with disperse dyes, the dyebath is slightly acidic to improve the stability of dye dispersion; however, the dyes are nonionic and therefore addition of electrolyte does not have a significant effect on improving dye substantivity and may in fact adversely affect the stability of dye dispersions. The substantivity of dispersed dyes may be improved to an extent by introducing various chemical groups into its structure or by modification of its structure. These methods include the following general approaches [58,119,152–154]:

(1) Incorporating polar groups such as hydroxyl and amino/substituted amino groups in dye structures.

(2) Inclusion of benzene or other groups, which can provide flatness or planarity to the molecule, thereby increasing dye substantivity.
(3) Increasing the number of conjugated double bonds.

12.1.4. Solubility

Disperse dyes have relatively low solubility (approximately 0.1–100 mg/L) in aqueous media under general dyeing conditions. Aqueous solubility at 80°C ranges from 0.2 to 100 mg/L. Solubility increases with temperature and is between 0.6 mg/L and 900 mg/L at 130°C, where the aqueous solubility of disperse dyes is some 3.5 times greater than that at 95°C [40]. Therefore, at 130°C a considerable amount of dye may be in solution. The increase in saturation value (S_f) of fiber because of an increase in application temperature can be attributed to an increase in aqueous solubility of the dye [159].

The adverse effects of many nonionic products on the stability of dye dispersion have been known for some time. Very severe aggregation of dye can affect the dyeing properties of polyester fibers. The *clouding point* is the temperature at which precipitation of solid matter in a fluid commences and phase separation of two phases gives the fluid a cloudy appearance. Dye aggregation can take place when the temperature of the dyebath is raised above the clouding point of auxiliary compound(s); especially in the presence of nonionic agents of relatively low clouding point [160]. The solubility of disperse dyes is also influenced by the following additional factors [161,162]:

- Crystal size and amount of dye molecule.
- Aggregation and agglomeration of dye.
- Changes in dye's crystal structure.
- Solubilizing effects of dispersing agents and other surfactants on dye.
- Concentration of dispersing agent.
- pH value.

Different crystal forms of a chemically identical disperse dye have been found to exhibit different saturation values on PET [163]. This is considered to be due to a difference in the solubility of different crystallographic forms of dye in water and in fiber [164]. Selected nonionic surface-active agents can be used to increase the solubility of disperse dyes in water. The use of nonionic surfactants results in the following benefits [165–167]:

- Increased migration, levelling, and fiber penetration.
- Increased rate of dyeing of relatively insoluble and more complex dyes. This is of advantage when disperse dyes of better than average wet fastness are applied to acetate and nylon.
- Reduced tendency for listing and ending in jig dyeing.

In addition, the solubility of disperse dyes is affected by temperature. The dissolved dye will diffuse into the fiber, but perhaps may form crystals first. These crystals are of the same system as the starting particles (during crystal growth), but display a new structure (changed or modified), or form only as the liquor cools [39]. The solubility of disperse dyes is also of importance when dye stripping is desired. Under high-temperature conditions (125–135°C), dye stripping is quicker, but depends on the water solubility of the dyes [4].

12.2. System parameters

12.2.1. Temperature

Polyester fibers arc usually dyed in the region of 125–135°C because of the low diffusion rate of disperse dyes at temperatures up to the boiling point [92]. Heat setting before dyeing changes the capacity of polyester to absorb dyes. The temperature of the dry heat-setting treatment affects the uptake of disperse dyes [70]. The effect of variation in temperature during heat treatment and dyeing on uniformity of dyeing is more severe and can result in a relatively large color difference between the differently treated fibers that are dyed in the same bath [4]. When dyeing polyester with disperse dyestuff, absorption first falls with an increase in heat-setting temperature and reaches a minimum at 160–180°C and then starts rising [32]. High heat-setting temperatures in the presence of water should be avoided because of the resulting fiber damage. Also, increasing steam pressure in the range used for twist-setting polyester yarns causes a gradual reduction in the quantity of dye taken up by the fiber over a given period.

During dyeing, the adsorption of disperse dyes onto PET fibers follows a Nernst isotherm obeying the simple partition mechanism described by the following relationship:

$$K = D_f/D_s = S_f/S_s,$$

where K is the partition coefficient, D_f and S_f represent the concentration of dye on fiber and the saturation value of dye on fiber, respectively, and D_s and S_s represent those in solution, respectively [88]. An increase in the application temperature decreases the partition coefficient K, but increases the saturation value (S_f) of disperse dye. The reduction in K that accompanies an increase in temperature can be attributed to the exothermic nature and the negative entropy changes involved in disperse dye adsorption together with an increase in solubility of the dye in the aqueous phase [110,117]. The increase in the dyeing rate, as the temperature rises, is associated externally with an increase in the water solubility of the dye and internally with an increase in the rate of diffusion of the dye in the fiber. High-temperature dyeing therefore gives much improved levelling action as well as rapid build-up of color.

12.2.2. Dye particle size

Disperse dyes are commonly milled in order to achieve the desired particle size and particle size distribution [71]. Since the dispersion of disperse dyes is thermodynamically unstable, the particle sizes tend to increase due to the decreased free energy of the system [22,95]. In addition, the enlargement of dye particles can be accelerated by desorption of dispersing agents from the surface of dye particles under dyeing conditions [168]. The larger the particle size, the greater the attraction energy between the dye particles [169]. Large particles are thus more likely to aggregate with each other, and dye dispersion gradually becomes unstable with increasing particle size. Therefore, the particle size of disperse dyes is closely related to their dyeing properties [170,171]. Owing to the range of particle sizes present in a typical dye dispersion and the higher solubility of the smaller particles, the dye solubility may become super-saturated with respect to the larger particles, which in turn will reduce the mean solubility of dispersion and thus reduce dye uptake. Such crystal growth, which can occur during cooling of the dyebath during high-temperature dyeing and shading operations, can be promoted by nonionic levelling agents. Anionic dispersants, however, stabilize a disperse dye dispersion against crystal growth [172]. The effect of adding appropriate dispersants in a disperse dyebath is to lower the limiting solubility of the dispersion and hence promote the optimum dye uptake by the substrate [153]. In order to achieve a desirable particle size and size distribution for optimal dyeing property, therefore, disperse dyes are usually milled in the presence of dispersing agents [58,173].

12.2.3. Dispersing agents

Dispersing agents can increase the aqueous solubility of disperse dyes in the dyebath and aid the production of uniformly dyed goods. However, excessive additions of dispersing agents to the dyebath should be avoided, as this would impair the degree of exhaustion of the liquor and the rubbing fastness of dyed material [8]. The main functions of dispersing agents are [22,160,161]

- to assist in reducing the particle size of disperse dyes;
- to facilitate reverse changes from the powder to dispersion; and
- to maintain uniform dispersion during dyeing.

The choice of dispersing agent is of great importance. Dispersing agents are mostly anionic products such as naphthalene sulphonic acid, cresol, and lignosulphonates [173]. Of these, naphthalene sulphonic acid is in extensive use. However, dispersing agents may cause staining of fabric, specky and unlevel dyeing, and have a reductive effect on dyes. These are also discharged in effluent with the residual dyeing liquor, increasing the Chemical Oxygen Demand (COD) and Biochemical Oxygen Demand (BOD) values of effluent and cause other environmental problems [111]. One of the main factors causing the aggregation of disperse dyes at high temperatures is the breaking away of dispersing agent from dye particles into the liquor.

12.2.4. Levelling agents

Levelling agents in use, other than the carriers, are ethoxylated products exhibiting nonionic character. These are added to the dyebath to obtain level dyeings. They function by reducing the strike rate of the dyes, which is achieved by retaining the dye in the dyebath by virtue of their solubilizing effect [167]. As the solubilizing effect of ethoxylated products is quite high, this results in reduced color yield, the extent of which depends upon the dye type and the quantity of levelling agent added to the dyebath. Among the levelling agents employed in disperse dyeing are nonionic ethylene oxides, which help improve the levelling properties of the dye by keeping the dye in dispersion, thus slowing down its exhaustion. These are soluble in water but their solubility decreases with increasing in temperature to the extent that at a certain temperature they become insoluble. The temperature where levelling agents become insoluble is termed the *cloud point*; therefore, it is important to select levelling agents that have cloud points above maximum dyeing temperature [61,156].

12.2.5. Liquor ratio

Liquor ratio is defined as the ratio of the amount of liquor in the dyebath to the amount of fabric (expressed by kg liquor/kg fabric). The use of low liquor ratios results in improved exhaustion levels [88] and generates higher color yields. However, if the ratio is too low, unlevelness may occur. The enhancement in the strength of dyeing, expressed by the K/S value of the dyed polyester substrates, as the liquor ratio decreases could be attributed to the greater swellability and accessibility of the polymer structure along with a greater availability of dye molecules and active ingredients in the vicinity of the so-called amorphous regions of the polyester polymer [53]. The control of liquor ratio, as a key parameter, is highly important as it strongly influences the amount of water, energy, and chemicals used at every stage of the textile processing where batch-wise operations are employed. Conventional methods of dyeing take place at liquor ratios between 10:1 and 30:1 for fabrics and between 10:1 and 15:1 for yarns. By comparison, the term 'short liquor ratio processing' is today taken to mean processing at liquor ratios in the range of 4:1 to 6:1. The most important parameters in short liquor dyeing, which should be controlled to generate level dyeing results, include [64,174]

- flow rate;
- contact number (defined by flow rate/liquor ratio); and
- concentration of dyestuffs and chemicals.

12.2.6. Dyebath pH

Some disperse dyes degrade during aqueous dyeing, giving rise to off-shade and pale dyeings, although in some cases such pH-induced changes are reversible [159]. It has been concluded that color yield obtained with disperse dyes decreases with increasing pH, and in some cases application at pH of 8–10 destroys the dye [175]. Dulling of shade of anthraquinone disperse dyes can sometimes occur due to the presence of metals in the dyebath [176]. Normally, aqueous dyeing is carried out in the pH range 4–4.5, the dyebath pH being adjusted using acetic acid or a suitable buffer system. When dyeing at high temperature, alkalinity in the dyebath should be avoided. Not only does alkalinity cause some loss in the yarn strength but certain disperse dyes – such as Dispersol Fast Yellow A (C.I. Disperse Yellow 3) and Duranol Blue Green B (C.I. Disperse Blue 7) – show decreased exhaustion under alkaline conditions or, as in the case of Dispersol Fast Yellow GR and Dispersol Fast Rubine BT, may even cause decomposition [70]. For these reasons, when dyeing at high or normal temperatures, the dyebath should be kept at pH of 5–6, for example with a buffer such as sodium or potassium dihydrogen phosphate. Chelating agents are used to control the shade stability of metal-sensitive anthraquinone dyes, and azo dyes are used for economical reasons [110]. It has been observed that certain azo dyes lose color at high pH in the presence of chelating agents. The color loss with sequestering agent is not a factor if dyebath pH is properly controlled in the range of 4–5. A pH of 5.5–6.5 is preferred in the dyeing of polyester/cotton blends with disperse dyes [77]. Also, polyester fibers can be degraded at high temperatures, especially if the process water is allowed to become alkaline. Dyebaths, therefore, should be buffered to a pH within the optimum range (pH 4.5–6) [4]. If shade corrections are required, for instance, if a grey shade gets greener after the addition of a further quantity of a rubine dye, the pH of the bath should be corrected first by adding a suitable buffer, such as a mixture of the acid sodium phosphates, prior to adjusting other parameters to bring the pH of the bath to 5.

12.2.7. Carriers

In spite of the many advantages of high-temperature aqueous dyeing, circumstances arise in which its use is undesirable, either because of the presence of some other fiber that is unstable to the severe conditions, or because of the lack of suitable equipment, or because of some other factor that may have arisen within the complex diversity of the textile industry [23,152]. Carriers have been used to resolve such issues. Carriers are often referred to as plasticizing agents, as they promote dye migration and transfer to fiber through the loosening of inter-chain forces between the fiber molecules, resulting in level and satisfactory dyeing, as well as prints [12,63]. Carriers permit easier movement of the polymer chain molecules and promote free volume availability in the fiber. The associated increase in segmental mobility of polymer chains may be expected to increase the rate of dye diffusion and decrease T_g of the fiber.

Polyester yarns and fabric typically can be treated in biphenyl/methylene chloride solutions at 21°C for 45 sec. This method is thought to utilize the rapid rate of diffusion of methylene chloride into polyester to promote the transport of carrier into the substrate prior to improving dyeability [129]. The use of carriers is intended to increase the exhaustion of dyes, particularly deep shades, improve the migration and diffusion of dyes, and assist the coverage of fiber irregularities. At the same time, however, carriers accelerate the strike rate of dyes in the initial phase of dyeing and can thus cause unlevelness [173]. However, an alternative mechanism for the action of water insoluble carriers could be due to the formation of a surface layer of carrier around the fiber in which the dye is very soluble and which aids the rapid movement of dye into the fiber because of the intimate contact between dye and fiber [177]. Some carriers cause fiber shrinkage in the dyeing of polyester fibers with disperse dyes [178]. For instance, fabrics woven from staple fibers crimp-set at temperatures around 130–135°C have been reported to experience up to 8% shrinkage [4]. In addition, toxicity, residual odors, and the effect of carrier residues on light fastness of dyed substrates must be considered, also the relative effectiveness of the carrier with the classes of dye that are to be employed. Many carriers have a deleterious effect on the light fastness of disperse dyeing, but the effect can in most cases be reduced by treatment with hot air between 150°C and 180°C for at least 30 sec [136]. When dyeing is carried out in totally enclosed systems, excessive quantities of any carrier that has a solubilizing effect on the dye should be avoided, otherwise poor color yields will be obtained [167].

12.2.8. Exhaustion

A typical exhaust dyeing application sequence for polyester can be divided into three main phases of the process: (1) adsorption phase (heating), (2) diffusion phase (at high temperature), and (3) reduction clearing phase as shown in Figure 15. Dye exhaustion should be as early and as complete as possible to satisfy the demand of short fixation times and to prevent subsequent exhaustion on cooling or even recrystallization and agglomeration of recrystallized dye particles on the fiber surface [135]. Research has shown that some disperse dyes exhaust differently in combinations than separately, though this finding is not commonly agreed upon in the industry [179]. When dyes exhaust differently in

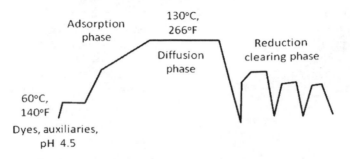

Figure 15. Phases of exhaust dyeing of polyester.

combination than separately, they may be mutually affecting each other because of a reaction in the dyebath, at the fiber–dyebath interface, or inside the fiber. A possible influencing factor may be the larger amount of dispersing agents present in a combination shade. This effect is less distinct in the presence of larger amounts of two dyes that cause a displacement effect at the fiber surface [132]. If only one dye is used, determining critical exhaustion is easy through a laboratory procedure. However, very few shades are matched with one dye, and dye manufactures therefore group dyes according to their individual rates of exhaustion for use in combination to ensure level and reproducible dyeings. There is a significant exhaustion temperature during the dyeing phase, which corresponds to a temperature where polymer chains achieve mobility. This mobility allows dye penetration, and this significant temperature is influenced by the size and steric configuration of dye molecules. Concentration must be considered when determining the exhaustion temperature of a dye because as concentration increases, exhaustion temperature also increases [53]. Up until this temperature is reached, the rate-of-rise can be rapid because no significant exhaustion takes place. The ability to increase temperature minimizes time and energy that was spent with a lower and constant rate of rise. Minimizing time is also important to limit the exposure time of the dyes to external forces, which can break up dye dispersion [180].

The exhaustion of dye is facilitated by the following factors [132,179,181,182]:

- Improved absorbency of the substrate.
- Increased dwell time in the impregnation bath, which is machinery-related.
- Any operation that would physically push the chemicals into the substrate.
- Providing energy to increase the diffusion and penetration process.
- Constant bath concentration gradient by dozing of chemicals from the stock liquor.

12.2.9. Levelness

Many factors are involved in levelling property of dyed polyester substrates. These factors include small variations in spinning and drawing conditions during fabric manufacturing, the solubility of disperse dyes, pH, and type and amount of auxiliaries such as carrier, levelling agent, dispersing agent, emulsifier, and antifoaming agent, etc. Carriers essentially shift the fixation to lower temperatures without radically changing the rate of adsorption [4,183]. The levelness issues arise in the strike stage of dyeing processes, when dye is being transferred from the dispersion to the fibers, and their elimination depends on the effectiveness of the redistribution of dye during the subsequent levelling stage, which is effectively due to dye migration. The best conditions for level dyeing exist in polyester dyeing when the dyes are uniformly distributed over the entire body of material as quickly as possible after the beginning of the dyeing process and then exhaust on tone [135]. Thus, the levelness of the polyester dyeing depends on [103,144,184,185]

- stability of the dye dispersion;
- solubility of the dyes;
- contacts between the liquor and fiber;
- dyeing temperature and time; and
- diffusion rate of the dye into the fiber.

12.2.10. Migration

Disperse dyes show thermal migration, i.e. migration of dye from core to surface during drying of dyed goods at 170°C or above [186]. Migration of disperse dyes of high molecular weight that have high sublimation fastness is pronounced. Whenever thermal migration poses a low wash fastness problem, the remedy is to post-scour the dyed material followed by drying at 140°C or below [187]. During storing of the dyed material, some disperse dyes have the tendency to be desorbed from polyester and be reabsorbed in the layer of oil or finishing agent, particularly under the influence of heat, humidity, and pressure. As the dye migrates from the core to the surface of the material, the fastness to light, sublimation, and wet treatments may be adversely affected. In general, nonionic emulsifiers and finishing agents are responsible for thermal migration.

12.2.11. Fastness

Disperse dyed polyester substrates show remarkably good fastness properties. During the 1950s and 1960s, the development of high-temperature processing methods placed sever demands upon disperse dye technology in terms of fastness [29]. Disperse dyes are considered to be fully fixed when optimal depth of shade and fastness is reached and subsequent hot air treatment causes no further change in shade [162]. Some researchers have highlighted the importance of establishing correlations between disperse dyes molecular structure and their fastness properties. General tendencies have been observed, for example, switching the halogen group of 2-bromo-4,6-dinitrophenyl diazo components with a cyano or nitro functional group was noted to improve wet fastness [188]. During the development of new disperse dyes for polyester fiber it was recognized that sublimation and migration fastness properties seem to be contradictory. Sublimation fastness is probably the most important fastness property of dyed polyester after their light fastness. Migration occurs during drying or heat treatment of polyester fibers containing spinning oils, nonionic surfactants, spin finishes, or other auxiliaries, which reduces the color fastness of the dyed textile goods [186]. Also, the thermomigration phenomenon, which occurs after heating dyed polyester, has been shown to also impair the crock, water, wash, and perspiration fastness. In addition, time migration of dye can occur after extensive storage of treated textiles without major exposure to heat [69]. Many piece-dyed goods are heat-set before dyeing, but some have to be set after dyeing. Even a very slight stain on white or light colored fibers or yarns present in a patterned fabric can produce a noticeable change in its overall appearance. In addition, heat setting after dyeing affects thermomigration of dyes and causes the dye to no longer be fixed inside the fiber. This problem will drastically affect the fastness properties of the fabric [56]. Thus, it is necessary to take care in the selection of dyes for such end uses to ensure that the fastness of the dyed yarns or fibers will be adequate [189].

12.3. Problems originating in equipments

The use of high-quality machinery on the factory floor greatly improves the frequency of 'Right First Time' (RFT) dyeing. Additional equipment can be purchased to monitor dyeing efficiency, but in general a good understanding of dye chemistry and good shade-matching equipment has been suggested to be sufficient. On the factory floor, machines that are automated will improve the dyeing process by maintaining correct time, temperature, and pH. It is important to select appropriate equipment for the given process based on end-use requirements of products and to ensure good maintenance and performance. Improper cleaning action on equipments gives rise to problems with poor wash, crock, and light fastness and thus obtaining the appropriate shade on the fabric becomes difficult [134,190–192].

The choice of dyeing equipment utilized in textile dyeing influences the substantivity of the dye. Yarns can be dyed in a package dyeing machine, rope goods in a jet or beck, and flat goods in a jig. The important factors to consider in the selection of machinery are the flow patterns of the fabric and solution, to determine the effects on substantivity. The higher the interactions between the goods and dye solution, the higher is the substantivity. Having a dynamic solution with a high flow rate allows the hydrodynamic boundary layer to be compensated for and increases the affinity of the dye.

12.3.1. Dyeing machines

Dyeing machines may be broadly classified into two types [115,190,193,194]:

1. Batchwise dyeing machinery allowing processing in rope-form (beck, jet) or in open width (jig).
2. Continuous dyeing machinery aimed for high-volume production.

The following are the basic requirements of dyeing machinery [115,190,193,194]:

- Providing sufficient movement for the liquor to penetrate uniformly into the textile materials.
- The liquor movement should not be so vigorous that the material is damaged, entangled, or felted.
- The material used for the construction must be stable during prolonged boiling with acidic or alkaline solutions.
- The heating arrangements of the machine should be able to maintain uniform temperature throughout the liquor.
- There should be a provision for addition of concentrated dye solution in such a way that it is well diluted before it encounters the textile materials.
- All moving parts and attachments should be protected against the corrosive action of steam and acidic fumes.

Atmospheric *beck* (or *winch*) can be used for dyeing at temperatures up to 100°C. Pressurized beck is used for dyeing at temperatures higher than 100°C. The greatest advantages of becks are their simplicity, versatility, and relatively low price. Beck dyeing machines subject fabrics to relatively low lengthwise tension and encourage the development of yarn crimp and fabric bulk. However, becks tend to use large amounts of water, chemicals, and energy. Becks can also cause abrasion, creasing, and distortion of some fabrics. The deficiencies of beck are listed as follows [27,29,30,115,135,157,192]:

- Formation of running creases during dyeing, which may not be removed even by stentering.
- Considerable longitudinal tension exerted on the goods resulting in elongation and deformation.
- Rapid heating and cooling promoting deformation.
- Dimensional stability not ensured even if the material is heat-set before dyeing.
- Long liquor ratio hampering economy of dyeing.
- Risks of entanglement of fabric rope.
- Difficulty in maintaining uniform temperature throughout the dyebath.

The most commonly used production dyeing machine is the *jet*. The jet dyeing system utilizes a pressurized water jet to move fabric through the machine, and a pump to circulate the jet and dye liquor together. The jet dyeing system was introduced as a way to have both dynamic goods and liquor during the dyeing process. This enables the highest amount of dye and fiber interactions, which allows for the highest standard potential of the dye to be achieved. Also, jet machines can be used at low liquor ratios, making them the choice of machinery to carry out reactive dyeing and achieve the highest affinity from the dye. Jet dyeing machines provide the following advantages compared to atmospheric becks for dyeing fabrics made from texturized polyester [27,29,30,115,135,157,192]:

- Vigorous agitation of fabric and dye formulation in the cloth tube increases the dyeing rate and the uniformity.
- Rapid circulation of fabric through the machine minimizes creasing because the fabric is not held in any one configuration for very long.
- Lengthwise tension on the fabric is low, so the fabric develops bulk and fullness of handle.
- Dyeing at high temperature of about 130°C gives rapid dyeing, improved dye utilization, improved fastness properties, and makes possible the elimination of carriers required when dyeing at lower temperatures.
- The lower liquor ratio used in jet dyeing allows shorter dye cycles and reduces the consumption of chemicals and energy.

Some disadvantages of jet dyeing machines compared to becks are as follows [27,29,30,115,135,157,192]:

- Capital and maintenance costs are relatively higher.
- Limited accessibility makes cleaning between dyeings and sampling for color during the dye cycle difficult.
- The jet action tends to make formulations foam in partially flooded jet machines.
- The jet action may damage the surface of certain types of fabrics.

The main components of a *jig* (or *jigger*) machine consist of a trough for the dye or chemical formulation and a fabric roll. The fabric is run from a roll on one side of the machine through the formulation in the trough and wound on a roll on the opposite side. When the second roll is full, the drive is reversed and the fabric is transferred through the formulation back to the first roll. A jig machine is normally used for dyeing under atmospheric pressure although pressurized, high temperature jigs have been also made. Covering the top of a jig minimizes heat loss to the atmosphere, keeps the temperature uniform on all parts of the fabric, and minimizes exposure of the formulation to air. Minimizing exposure to air is most important when using sulphur and vat dyes since these dyes can be oxidized by atmospheric oxygen. Maximum batch size on a jig may be up to several thousand meters of fabric. Jigs exert considerable lengthwise tension on the fabric and are therefore more suitable for woven than for knitted fabrics. Since the fabric is handled in open width, a jig is very suitable for fabrics that crease if dyed in rope form.

Yarn is dyed in hank form or in the form of packages, namely cones, cheeses, or bobbins. All yarn packages for dyeing must ensure the following [27,29,30,115,135,157,192]:

- Adequate stability during handling, loading, and unloading of yarn in the machine spindle.
- They withstand the change in temperature of the flowing liquor and pressure during reversal of flow.
- All packages accommodate a small amount of shrinkage in some yarn and the effect of fiber swelling when wetted out.

- The wound yarn is not trapped, otherwise the back-winding of the packages will be disturbed.
- Fairly sufficient resistance to liquor flow at different temperatures and pressures both within and between the packages.

The principles of *beam* dyeing are essentially identical to those of package dyeing. Either yarn or fabric can be beam-dyed. The fabric or yarn is wound on a perforated beam. A beam machine can be designed to hold a single beam or multiple beams in a batch. Beam dyeing of warps is practical in producing patterned fabrics where the warp yarn will be of one color and the filling will be of another color. The dyeing faults, which commonly occur in beam-dyed materials, include the following [27,29,30,115,135,157,192]:

1. Gross unevenness near the edge mostly because of slack winding.
2. Uneven or light patches because of air bubbles.
3. Variation near the edge on inside layers because of misjudged overlap of the steel collar.
4. Variation in shade between the inside and outside layers. The inner layers are relatively more deeply dyed than the outside layers as the concentration of dye diminishes when the dye liquor passes through the roll, especially when high-affinity dyes are employed.

In *hank* dyeing (also called *skein* dyeing), skeins of yarn are mounted on a carrier that has rods (sticks) at the top and bottom to hold the skeins. The skeins are suspended in the dyeing machine and dye liquor is gently circulated around the hanging skeins. Perforated plates can be used at the top and bottom of the machine to help provide a uniform flow of the dye liquor. Alternatively, the dye liquor can be pumped through perforations in the sticks so that it cascades down over the hanging skeins. Hank dyeing produces good bulk in the yarn because of the low tension on the yarn in the dyebath. The method is used mainly for bulky yarns like acrylics and woollens for knitted outerwear and hand-knitting. Woollen carpet yarn is sometimes skein-dyed. Skein dyeing uses a high liquor ratio and a lot of energy. Uniform dyeing is also difficult to achieve in a skein dyeing machine. Slow winding and back-winding requirements of the process make it labor-intensive. Package dyeing has thus replaced most of hank dyeing even though the yarn bulkiness achieved in skein dyeing is usually not matched in package dyeing. The following are the advantages of package dyeing over hank dyeing [27,29,30,115,135,157,192]:

- Elimination of hank reeling.
- Reduced waste.
- Faster back winding.
- More controlled dyeing with better levelness and fastness.
- Lower liquor ratio causing savings in water, effluent, energy, dyes, and chemicals.
- Lesser floor space.
- Less labor-intensive.
- High-temperature dyeing and rapid drying possible.
- Larger control on the process makes automation easier.

12.3.2. Issues related to other machinery

Many dyers describe creasing as the most frequently encountered problem on the continuous dyeing range. A *padder* is usually employed to apply the required chemicals to the substrate.

Two major problems attributed to difficulties with *padding* are shading, either side-to-side or side-to-center, and dye spots.

Availability of a satisfactory steam supply is critical. Steam that is too hot and dry will cause drying of the fabric and prevent reduction. Other problems originating at the *steamer* include water spots and loss of depth or change of shade.

Washing is necessary not only after dyeing but also after scouring, bleaching, mercerization, etc. With the introduction of polyester and the use of wider fabrics, it became almost impossible to avoid problems due to creasing in rope processing machinery. *Wash boxes* serve the purpose of initial rinse to remove the chemicals applied in the chemical pad and improve the final fastness characteristics of the fabric by removing unfixed dyes. The efficiency of wash boxes to produce these results with a minimal consumption of water has become of considerable importance in recent years. Recently, open-width wash boxes have gradually replaced rope washers. Some of the merits of rope washers are as follows [27,29,30,115,135,157,192]:

- Flexibility with fabric width up to a certain limit.
- Easy movement of fabric over long distances.
- Higher running speeds.
- Less storage space.
- Less floor-space for the washer.
- Less capital and maintenance costs.
- Less tension on the fabric.
- Better relaxation of fabric resulting in better handling.
- Savings of water, time, and labor.

The following are the disadvantages of rope washing [27,29,30,115,135,157,192]:

- Unsuitable for certain fabric types and weights.
- Need for additional opening stage after washing.
- Fabric structure is prone to distortion or creasing.
- Lightweight fabrics are prone to tangling.

The purpose of the *predryer* is to remove a sufficient amount of water from the padded fabric to reduce the migration capability of the dye molecules. Predrying is not intended to stop the migration of dye particles but to control it so that dye molecules will be evenly distributed throughout the fabric. However, dye migration is a major problem in this step, which needs to be controlled by temperature and time. In the case of polyester, the appropriate condition used commonly for heat presetting is 180°C for 30 sec.; however, it is recommended that appropriate temperatures and time should be decided after testing the breaking stress of fiber. The advanced types of infrared predryers are designed to cool quickly and avoid the swing away issue during machine stoppage. However, general predryers are not equipped with infrared heaters. In order to control the rate of water evaporation, the passage length inside the drier can be increased and the temperature gradients at different sections of the drier are kept minimal. If the temperature gradients were high, the mobile water would tend to evaporate faster and facilitate migration. Drying is especially an important process for both dyed package and dyed yarns. The yarn shipped to the knitter or weaver must contain the commercially acceptable amount of moisture, as any deficiency is a direct loss of profit. The moisture content of the yarn going from the dye house to the winding department must also be uniform, as variations lead to uneven pick up of wax, which in turn can give an uneven coefficient of friction and erratic knitting.

Insufficient drying leads to packages being shipped noticeably damp, which results in faulty dyeing and irate customers.

Dryers should also have sufficient pressure at the end of the range to completely dry the fabrics. Drying significantly improves the appearance of the dyed fabric, and reduces the residual moisture in the fabric. It also enhances the dye range efficiency by up to 20–25%. Consideration should be given to the weight of dyeings and the desired machinery's speed, temperature, and dwelling time. The stenter is the only drying machine that provides adjustment and control of fabric width during drying. Stenter performs several other functions, which include the following [27,29,30,115,135,157,192]:

(1)　Heat-setting of fabrics made of polyester and blends.
(2)　Application and fixation of several finishing agents.
(3)　Imparting particular mechanical finishes affecting appearance and feel, commonly known as stenter finish.

12.4. Problems originating from human factors

The faulty dyeing is directly or indirectly caused by reasons connected with human resources such as chemical operator, control operator, and dye shift operator or the communication failure among them. In the dyeing process, the following additional factors should be considered [83,107,123,162,194–196]:

12.4.1. Chemical storage

The aged and/or low-quality chemicals and dyestuff affect the dyeing property of the final goods. The chemical store should only carry the necessary inventory of chemicals and dyestuffs, and at any point of time, and if compounds become redundant; they need to be disposed of. There should be a periodic review on the inventory of stores and identification of redundant items. Before disposing of chemicals, the stores, however, verify whether any other departments would need any of these items. A written formatted procedure should be established to record the transactions to avoid essential items getting disposed of due to indiscretion. Basic precautions such as the following must be taken:

- Highly reactive chemicals need to be stored in closed compartments to avoid exposure to moisture.
- Some of the binders have a certain shelf life and need to be used up in time.
- Some chemicals such as hydrosulphite of soda and hydrogen peroxide need to be stored separately.
- Segregated weighing areas must be established to avoid incompatible commodities getting contaminated during weighing, e.g. weighing of sodium dithionite away from the weighing zone of dyes.

12.4.2. Equipment and maintenance

The success of dye house is closely linked with its engineering maintenance support, without which the operations can come to a grinding halt. The other aspect is the role played by individuals in supplying and maintaining the utilities. Only the needed tools and equipment maintenance kit should be stored in the department. It is a common experience to notice the maintenance tools and items remaining in the work place long after the maintenance is carried out. Such items mess up the place and could be a cause for accidents and also damage dyed products. Thus, it is necessary that the operators are trained with respect to the fundamentals of practice so that they consciously understand that the unit is under control irrespective of whether the unit is modern or employs older version machinery and that when things go wrong they would be in a better position to resolve problems.

12.4.3. Discipline and training

A knowledgeable and cooperative dyer and his team are the most important assets for any dye house. The most challenging part of the training is the acceptance of the principles and adopting the same as mandatory. This is not a one-time exercise and therefore the practice should become a habit. Any new recruit should be trained before he or she is assigned the work. The personnel should acquire the necessary attitudinal skills to achieve the work culture to function as a team complementing and supplementing each other. The discipline and training is an ongoing process, which would inculcate a stage of harmony and self-confidence in the work force. A well-maintained machine would work trouble-free, turn out good quality products, and give higher productivity. The men would be motivated to achieve high standards of performance. The program for the day for the machine and men would be clearly understood and carried out in the time span scheduled. The morale, enthusiasm, and discipline of the men would be high, and procrastination, a fundamental negative factor in a competitive work environment, should have no place.

12.4.4. Color matching communication

In the case of human communication, which involves interaction between two or more individuals, directly or in a chain, the interpretation of what is said and what is understood by different individuals is not likely to be the same. In one-to-one or group relationship, clarifications and explanations sort out the understanding better. However, in a chain communication, the clarification and/or explanation phase may not always exist, and at that stage the problems start. In a verbal situation, the understanding at the end of the chain could be totally distorted, each link contributing to the confusion.

For example, on a number of occasions, the supplier does not understand the user requirements of what constitutes the right fabric 'hand'. Metameric matchings, and shade tolerances are some of the very common causes for contention and much energy and time are wasted in sorting out these issues. Concurrence between the user and the vendor on the shade and acceptable tolerances thus need to be sorted out first.

12.4.5. Cleanliness for better motivation

Cleanliness of the dye house surroundings motivates operators to work better. It is a common observation that the cut ends of the fabric after stitching or detached end pieces at the end of an operation are thrown around on the floor near the vicinity of the stitching area and/or machine. Besides, cleaning the machinery and the surroundings is the responsibility of the concerned operator. He or she should keep his/her machine clean and in a constant state of readiness all the time.

13. Summary

This manuscript provides a review of possible problems caused by many factors in a dyeing process. It has been shown that dyeing as a process contains a considerable number of variables, and errors in any or all of them can produce dyeings that are not acceptable, with the result that expensive shadings have to be carried out. Many faults that arise in earlier stages of processing become clearly visible for the first time after dyeing and it is necessary for the dyer to learn to recognize their symptoms. Some of the common dyeing problems include dye spots, migration, uneven dyeing, staining, shading, off-shade colors, poor hand, and poor fastness. In general, most of these faults result from the following:

- Ingredients (fiber, yarn, and fabric).
- Pretreatments (desizing, scouring, bleaching, and neutralization).
- Dyestuff and auxiliary (quality, compatibility, and weighing).
- Dyebath (temperature, time, pH, and electrolyte).
- Equipments (control units, and cleanliness).
- Water quality (metal impurities, hardness, and control).
- Processes (dyeing, wash-off, drying, and reduction-clearing).
- Human factors (chemical/control/dye shift operator, inspector).

A categorized list of possible causes associated with faults in the dyeing of polyester can be found elsewhere [197]. The dyeing strategy depends on the quality of the dyed material as judged by fastness properties. Dyeing and physical properties can be improved with a decrease in overall costs together with ecological advantages. The textile industry, especially the dyeing and finishing sector, needs to adopt a more critical attitude by solving dyeing problems in order to establish the most logical methods of improving human performance.

Acknowledgement

Special thanks are due to Miss Jing Chen for her work on organizing the references and to Dr. Jeff Joines for coadvising Dr. Woo Sub Shim's PhD project entitled 'A Diagnostic Expert System for the Coloration of Polyester Materials', which was completed at North Carolina State University in 2009.

References

[1] K.-F. Au and M.-M. Wong, J. Asia Text. Apparel 6 (2006), pp. 23–27.
[2] F. Ayfi, *2003–2004 Handbook of Statistics on Man-Made/Synthetic Fibre/Yarn Industry. Part One, Fibre for Better Living*, Association of Synthetic Fibre Industry, Mumbai, India, 2004, p. 177.
[3] K.H. Hatch, *Textile Science*, West Publishing, University of Arizona, Tucson, AZ, 1993.
[4] R. Broadhurst, *Dyeing of polyster fibres*, in *The Dyeing of Synthetic-Polymer and Acetate Fibres*, D.M. Nunn, ed., Dyers Company Publications Trust, Bradford, West Yorkshire, UK, 1979.
[5] W.H. Carothers, J. Am. Chem. Soc. 51(8) (1929) pp. 2548–2559.
[6] H. Mark and G.S. Whitby, *Collected Papers of Wallace Hume Carothers on Polymerization*, Inter-Science, New York, NY, 1940.
[7] S.J. Kadolph and J. Sara, *Textiles*, 7th ed., Maxwell Macmillan International, New York, NY, 1993.
[8] H. Ludewig, *Polyester Fibres*, Wiley-Interscience, London, 1964.
[9] N.J. Roseland, *Man-Made Fiber Producers' Handbook*, Textile Economics Bureau, New York, NY, 1983.
[10] S.G. Hovenkamp and J.P. Munting, J. Polym. Sci. A-1 8 (1970) pp. 679–685.
[11] W. Ingamells and A. Yabani, J. Soc. Dyers Colour. 93 (1977) pp. 417–423.
[12] K. Brendth, Colourage 10 (1983) pp. 27–29.
[13] V.M. Irklei, Y.Y. Kleiner, O.S. Vavrinyuk and L.S. Gal'braikh, Fibre Chem. 37 (2005) pp. 447–451.
[14] J.R. Lewis, *From Fleece to Fabric: A Practical Guide to Spinning, Dyeing and Weaving*, Robert Hale, London, 1983.

[15] T. Manabe, Text. Res. J. 33 (1963), pp. 221–224.
[16] C. Miller, AIChE J. 50 (2004), pp. 898–905.
[17] M. Lewin and E.M. Pearce, *Handbook of Fiber Chemistry*, 2nd ed., Marcel Dekker, New York, NY, 1985.
[18] W.A.S. Perkins, Text. Chem. Color. 23(8) (1991) pp. 23–27.
[19] E. Hoboken, *Kirk-Othmer Encyclopedia of Chemical Technology*, 3rd ed., John Wiley & Sons, New York, NY, 1981.
[20] E.V. Hencken, *Textiles: Concepts and Principles*, Delmar, New York, NY, 1984.
[21] P. Grosberg and G. Iype, *Yarn Production: Theoretical Aspects*, Textile Institute International, Manchester, UK, 1999.
[22] S. Heimann, Rev. Prog. Color. Relat. Top. 11 (1981) pp. 1–8.
[23] B.S. Ashkenazi, Colourage 4 (1984) pp. 23–25.
[24] S.M. Burkinshaw and G.W. Collins, Dyes Pigm. 25 (1994) pp. 31–48.
[25] S.W. Henry and C.T. Gary, Text. Chem. Color. 13(10) (1981) pp. 24–27.
[26] W.J. Lee and J.P. Kim, J. Soc. Dyers Colour 116 (2000) pp. 345–348.
[27] S. Buchholz, *Dyeing and Finishing of Polyester Fibres and Polyester Fibre Blends BASF*, Ludwigshafen am Rhein, Germany, 1966.
[28] D. Farrington, J. Soc. Dyers Colour. 105(9) (1989) pp. 301–307.
[29] I. Holme, Text. Horiz. 6 (1992) pp. 29–31.
[30] O.S. Kenkyusha, *Dyeing and Finishing of Polyester Fiber*, Osaka Senken, Japan, 1978.
[31] L.J. Johnson, *Fractionation of fibers in surfactant solutions on a spinning disk separator*, PhD thesis. University of Wisconsin, Madison, WI, 1993.
[32] D.N. Marvin, J. Soc. Dyers Colour. 70(1) (1954) pp. 16–21.
[33] J. Radhakrishnan, U.P. Kanitkar and V.B. Gupta, J. Soc. Dyers Colour. 113(2) (1997) pp. 59–63.
[34] J. Bernstein, J. Phys. D: Appl. Phys. 26 (1993) pp. B66–B76.
[35] V.B. Gupta, A.K. Gupta, V.V.P. Rajan and N. Kasturia, Text. Res. J. 54(1) (1984) pp. 54–60.
[36] M.H. Rao, K.N. Rao, H.T. Lokhande and M.D. Teli, J. Appl. Polym. Sci. 33(8) (2003) pp. 2707–2714.
[37] V.F. Isakov, A.P. Andronova, É.M. Aizenshtein and T.M. Zhadenova, Fibre Chem. 17(5) (1985) pp. 334–336.
[38] C.M. Boussias, R.H. Peters and R.H. Still, J. Appl. Polym. Sci. 26(12) (2003) pp. 4125–4133.
[39] W. Biedermann, J. Soc. Dyers Colour. 88(9) (1972) pp. 329–332.
[40] C.L. Bird, J. Soc. Dyers Colour. 70(2) (1954) pp. 68–77.
[41] P.W. Jensen, J. Polym. Sci. 28 (1958) pp. 635–638.
[42] A. Hebeish, S.E. Shalaby and A.M. Bayazeed, J. Appl. Polym. Sci. 22(11) (1978) pp. 3335–3342.
[43] A. Hebeish, S.E. Shalaby and A.M. Bayazeed, J. Appl. Polym. Sci. 27(10) (1982) pp. 3683–3690.
[44] A. Hebeish, S.E. Shalaby and A.M. Bayazeed, J. Appl. Polym. Sci. 26(10) (1981) pp. 3245–3251.
[45] A. Hebeish, S.E. Shalaby and A.M. Bayazeed, J. Appl. Polym. Sci. 26(10) (1981) pp. 3253–3269.
[46] A. Hebeish, S.E. Shalaby and A.M. Bayazeed, J. Appl. Polym. Sci. 27(1) (1982) pp. 197–209.
[47] M. Okoniewski, J. Sójka-Ledakowicz and S. Ledakowicz, J. Appl. Polym. Sci. 35(5) (1988) pp. 1241–1249.
[48] X. Ramis and J.M. Salla, J. Polym. Sci. B: Pol. Phys. 37(8) (1999) pp. 751–768.
[49] R. Huisman and H.M. Heuvel, J. Appl. Polym. Sci. 22(4) (1978) pp. 943–965.
[50] C.S. Krishana, Colourage 22 (1981) pp. 3–10.
[51] M. Mitsuishi, Y. Naruoka, M. Shimizu, K. Hamada and T. Ishiwatai, J. Soc. Dyers Colour. 112(11) (1996) pp. 333–335.
[52] J.S. Koh, I.S. Kim, S.S. Kim, W.S. Shim, J.P. Kim, S.Y. Kwak, S.W. Chun and Y.K. Kwon, J. Appl. Polym. Sci. 91(6) (2004) pp. 3481–3488.
[53] N.A. Ibrahim, M.A. Youssef, M.H. Helal and M.F. Shaaban, J. Appl. Polym. Sci. 89(13) (2003) pp. 3563–3573.
[54] A. Arcoria, A. Cerniani, R. De Giorgi, M.L. Longo and R.M. Toscano, Dyes Pigm. 11(4) (1989) pp. 269–276.
[55] W. Ingamells and A.M. Yabani, J. Appl. Polym. Sci. 22(6) (1978) pp. 1583–1592.
[56] P. Richter, Int. Text. Bull.: Dyeing/Print./Finish. 3 (1990) pp. 12–15.

[57] O. Glenz, W. Beckmann and W. Wunder, J. Soc. Dyers Colour. 75(3) (1959) pp. 141–147.

[58] J.R. Aspland, Text. Chem. Color. 24(12) (1992) pp. 18–23.

[59] T.L. Dawson and J.C. Todd, J. Soc. Dyers Colour. 95(12) (1979) pp. 417–426.

[60] K.V. Datye, Colourage Annu. (1997) pp. 121–128.

[61] M.L. Gulrajani and D. Dara, Colourage 32(13) (1985) pp. 25–26.

[62] D.P. Hallada, Am. Dyest. Rep. 50 (1961) pp. 445–450.

[63] A.A. El-Halwagy, J. Text. Assoc. 63 (2002) pp. 33–37.

[64] C. Oschatz, U. Ruf and F. Somm, J. Soc. Dyers Colour. 92(3) (1976) pp. 73–84.

[65] A.L. Simal and J.P. Bell, J. Appl. Polym. Sci. 30(3) (1985) pp. 1195–1209.

[66] C.M. Carr, *Chemistry of the Textiles Industry*, Blackie Academic and Professional, Chapman & Hall, Cambridge, UK, 1995.

[67] Y. Murase and A. Nagai, *Melt spinning*, in *Advanced Fiber Spinning Technology*, T. Nakajima, ed., Woodhead, Cambridge, UK, 1994.

[68] J.N. Etters, Text. Chem. Color. 4(6) (1972) pp. 160–164.

[69] B. Goossens, Melliand (English ed.) 14 (1985) pp. 93–100.

[70] G.K. Uranl, *The Dyeing of Polyester Fibres*, Imperial Chemical Industries, London, UK, 1964.

[71] S.R. Shukla and M.R. Mathur, J. Soc. Dyers Colour. 113(5–6) (1997) pp. 178–181.

[72] S.R. Shukla and M.R. Mathur, J. Appl. Polym. Sci. 75(9) (2000) pp. 1097–1102.

[73] T.-S. Choi, Y. Shimizu, H. Shirai and K. Hamada, Dyes Pigm. 50(1) (2001) pp. 55–65.

[74] D. Chidambaram, R. Venkatraj and P. Manisankar, J. Appl. Polym. Sci. 87(9) (2003), pp. 1500–1510.

[75] J.J. Lee, N.K. Han, W.J. Lee, J.H. Choi and J.P. Kim, Color. Technol. 118(4) (2005) pp. 154–158.

[76] J.S. Lee, D.M. Shin, H.W. Jung and J.C. Hyun, J. Non-Newtonian Fluid Mech. W. 130(2–3) (2005) pp. 110–116.

[77] G.R. Turner, Text. Chem. Color. 21(8) (1989) pp. 23–25.

[78] A.N Derbyshire, J. Soc. Dyers Colour. 90(8) (1974) pp. 273–280.

[79] H.-G. Kilian and M. Pietralla, *Transitions in Oligomer and Polymer Systems*, Steinkopff, Darmstadt, Germany, 1994.

[80] V.B. Gupta and J. Amirtharaj, Text. Res. J. 46 (1976) pp. 785–790.

[81] W.R. Railey, *The effect of disperse dyeing in the alkaline pH range on the reduction of oligomer deposition during package dyeing of textured polyester*, thesis. Institute of Textile Technology, Charlottesville, VA, 1997.

[82] V.S. Salvin, Am. Dyest. Rep. 54 (1965) pp. 272–278.

[83] M.D. Potter and B.P. Corbman, *Textiles; Fiber to Fabric*, 4th ed., McGraw-Hill, New York, NY, 1967.

[84] N. Balasubramanian, Colourage 6 (1997) pp. 15–26.

[85] M. Dohmyou, Y. Shimizu and M. Kimura, J. Soc. Dyers Colour. 106(12) (1990) pp. 395–397.

[86] L.A. Graham and C.A. Suratt, Am. Dyest. Rep. 67 (1978) pp. 36–39.

[87] I.A. Naik, Melliand (English ed.) 14 (1985) pp. 127–131.

[88] B. Prede, *Package dyeing, Wira collected papers*, Wira Rep. 158, Wira, Leeds, UK, 1971.

[89] K.V. Masrani and J.L. Handu, Colourage 9 (1984) pp. 15–19.

[90] D.M. Lewis and K.A. McIlroy, Rev. Prog. Color. 27(1) (1997) pp. 5–17.

[91] R.B. Love and A. Robson, J. Soc. Dyers Colour. 89(12) (1978) pp. 514–520.

[92] S.M. Burkinshaw, *Chemical Principles of Synthetic Fibre Dyeing*, Blackie Academic & Professional, Glasgow, UK, 1995.

[93] J. Lin, C. Winkelmann, S.D. Worley, J. Kim, C.-I. Wei, U. Cho, R.M. Broughton, J.I. Santiago and J.F. Williams, J. Appl. Polym. Sci. 85(1) (2002) pp. 177–182.

[94] J. Rivlin, *Introduction to the Dyeing of Textile Fibers*, Philadelphia College of Textiles and Science, Philadelphia, PA, 1982.

[95] H. Leube, Text. Chem. Color. 10 (1978) pp. 39–46.

[96] S. Li, R. Shamey and C. Xu, Color. Tech. 125(5) (2009) pp. 296–303.

[97] W.R. Pomfret, Text. Mon. 10 (1971) pp. 67–71.

[98] P. Bajaj, A. Sengupta and K. Sen, Text. Res. J. 50(10) (1980) pp. 610–617.

[99] S.M. Burkinshaw and A.E. Willis, Dyes Pigm. 34(3) (1997) pp. 243–253.

[100] A. Bayazeed, Am. Dyest. Rep. 74 (1985) pp. 38–40.

[101] V.M. Kohli, *Effect of winding parameters on density of wound packages,* in *Proceedings of 2nd Annual Technology Conference*, Bombay Textile Research Association, Mumbai, India, 1967.

[102] K.P. Ramachandran, *Winding, Silver Jubilee Monograph Series*, Bombay Textile Research Association, Mumbai, India, 1981.

[103] C.M. Pastore and P. Kiekens, *Surface Characteristics of Fibers and Textiles,* Surfactant Science Series, Marcel Dekker, New York, NY, 2001.

[104] X. Zhao, R.H Wardman and R. Shamey, Color. Technol. 122 (2006) pp. 110–114.

[105] X. Zhao, R. Shamey and R.H. Wardman, Res. J. Text. Apparel 9(3) (2005) pp. 64–70.

[106] I. Holme, Int. Dyer 192(8) (2007) pp. 8–9.

[107] T.L. Vigo, *Textile Processing and Properties-Preparation, Dyeing, Finishing and Performance*, Elsevier Science B.V., Amsterdam, The Netherlands, 1994.

[108] J.B. Goldberg, *Fabric Defects*, 1st ed., McGraw-Hill, New York, NY, 1950.

[109] M. Masao, *Advanced Fiber Spinning Technology*, Woodhead, Cambridge, UK, 1997.

[110] W.C. Ingamells, *The influence of fibre structure on dyeing*, in *The Theory of Coloration of Textiles*, 2nd ed., A. Johnson, ed., Dyers Co. Pub. Trust, Bradford, UK, 1990.

[111] S.Y. Lin, Text. Chem. Color. 13(11) (1981) pp. 24–28.

[112] E.V. Martin and C.J. Kibler, *Polyesters of 1,4 cyclohexanedimethanol*, in *Science and Technology of Man-Made Fibers, Vol. 3*, H.F. Mark, S.M. Atlas, E. Cernia, eds., Interscience, New York, NY, 1967.

[113] Q.H. Chen, K.F. Au, C.W.M. Yuen and K.W. Yeung, Text. Res. J. 73(5) (2003) pp. 421–426.

[114] M.D. Falick, *Knitting in America*, Artisan, New York, NY, 1997.

[115] S.V. Kulkarni, *Textile Dyeing Operations Chemistry, Equipment, Procedures, and Environmental Aspects*, Noyes, Berkshire, UK, 1986.

[116] R.T. Norris and A. Ward, J. Soc. Dyers Colour. 89(6) (1973) pp. 197–202.

[117] A. Johnels, *The dyeing of yarn package*, PhD thesis. University of Goteborg, Goteborg, Sweden, 1967.

[118] R.S. Bhagwat, Colourage 38(1) (1992) pp. 57–66.

[119] A.K.R. Choudhury, *Textile Preparation and Dyeing*, Science Publishers, Enfield, New Hampshire, USA, 2006.

[120] F.J.C. Fité, Text. Res. J. 65(6) (1995) pp. 362–368.

[121] U. Denter, S. Dugal and E. Schollmeyer, Melliand (English ed.) 14 (1985) pp. 177–178.

[122] L. Heinz and S. Klaus, Melliand (English ed.) 14 (1985) pp. 12–19.

[123] H.U. Schmidlin, *Preparation and Dyeing of Synthetic Fibres*, Reinhold, New York, NY, 1963.

[124] N. Awad and J. Hauser, Text. Chem. Color. 13(9) (1981) pp. 28–32.

[125] J. Interox, *A Bleachers Handbook*, Interox America, Houston, TX, 1980.

[126] S. Niu and T. Wakida, Text. Res. J. 63 (1993) pp. 346–350.

[127] J. Carbonell, J.H. Heetjans and M. Angehrn, Melliand (English ed.) 14 (1985), pp. 74–85.

[128] K. Datye and A. Vaidya, *Chemical Processing of Synthetic Fibers and Blends*, John Wiley and Sons, New York, NY, 1984.

[129] B. Smith and J. Rucker, Am. Dyest. Rep. 76 (1987) pp. 68–73.

[130] B.-R. Lim, H.-Y. Hu, K.-H. Ahn and K. Fujie, Water Sci. Technol. 49(5–6) (2004) pp. 137–143.

[131] F.U. Drady, *Finishing of Polyester-Fibres,* special ed. 16, Bayer Farben Revue, Germany 1979.

[132] P. Richter, Melliand Texilberichte (English ed.) 12 (1983) pp. 336–341.

[133] C. Tischbein, J. Soc. Dyers Colour. 105(12) (1989) pp. 431–437.

[134] L.R. Smith and O.E. Melton, Am. Dyest. Rep. 14(5) (1982) pp. 38–42.

[135] H. Tiefenbacher, Melliand (English ed.) 14 (1985) pp. 242–248.

[136] W.C. Wilcoxson, Am. Dyest. Rep. 73(5) (1984) pp. 16–21.

[137] I.-S. Kim, K. Kono and T. Talagishi, Text. Res. J. 67 (1997) pp. 555–562.

[138] W.B. Achwal, Colourage 46(32–33) (1999) pp. 568–571.

[139] A. Datyner, *Surfactant in Textile Processing, Vol. 14*, Marcel Dekker, New York, NY, 1983.

[140] R.K. Prud'homme and S.A. Khan, *Foams: Theory, Measurements, and Applications*, Marcel Dekker, New York, NY, 1995.

[141] R.A.F. Moore and H.-D. Weigmann, Text. Chem. Color. 13(3) (1981) pp. 34–35.

[142] I. Slack, Can. Text. J. 7 (1979) pp. 46–51.

[143] R.F. Hyde, G. Ashton, G. Thompson and K.A. Stanley, Colourage Annu. (1997) pp. 67–76.

[144] S. Claude, Am. Dyest. Rep. 10 (1979) pp. 42–46.

[145] R. Detchewa, I. Keray, M. Duschewa and S. Stojanov, Textilveredlung 12 (1977) pp. 26–31.

[146] F. Andrew, Am. Dyest. Rep. 67 (1978) pp. 48–50.

[147] W. Biedermann, J. Soc. Dyers Colour. 88 (1972) pp. 329–332.

[148] S.K. Pal, R.S. Gandhi and V.K. Kothari, Text. Res. J. 66 (1996) pp. 770–776.

[149] M.J. Drews and C. Jordan, Text. Chem. Color. 30(6) (1998) pp. 13–20.

[150] J. Trfes, Text. Chem. Color. 11 (1979) pp. 20–25.

[151] H.H. Summer, J. Soc. Dyers Colour. 92 (1976) pp. 84–99.

[152] D. Balmforth, C.A. Bowers, J.W. Bullington, T.H. Guion, and T.S. Roberts, J. Soc. Dyers Colour. 82 (1966) pp. 405–409.

[153] H. Braun, Rev. Prog. Color. 13 (1983) pp. 62–71.

[154] C.H. Giles, Text. Res. J. 31 (1961) pp. 141–151.

[155] R. Mittal, *Application of Disperse Dyes*, Ahmedabad, Textile Industry's Research Association, India, 1986.

[156] E. Abrahart, *Dyes and Their Intermediate*, 2nd ed., Chemical Publishing, New York, NY, 1977.

[157] R. Gupte and M. Nayak, Colourage 22 (1985) pp. 11–14.

[158] A.L.N. Rao, Colourage 2 (1990) pp. 13–17.

[159] H. Brody, *Synthetic Fibre Materials, Polymer Science and Technology Series*, Longman Scientific and Technical, Harlow, Essex, UK, 1994.

[160] A.N. Derbyshire, W.P. Mills and J. Shore, J. Soc. Dyers Colour. 88 (1972) pp. 389–394.

[161] J. Odvarka and H. Schejbalova, J. Soc. Dyers Colour. 110 (1994) pp. 30–34.

[162] C. Thoma, Am. Dyest. Rep. 67 (1978) pp. 52–61.

[163] V.A. Shenai and M.C. Sadhu, J. Appl. Polym. Sci. 20 (1976) pp. 3141–3154.

[164] W. Biedermann, J. Soc. Dyers Colour. 87 (1971) pp. 105–111.

[165] M.R. Bhatt and L.N. Chaturvedi, Colourage 3 (1986) pp. 29–32.

[166] E.C. Ibe, J. Appl. Polym. Sci. 14 (1970) pp. 837–846.

[167] A. Murray and K. Mortimer, Rev. Prog. Color. Rel. Top. 2 (1971) pp. 67–71.

[168] E. Schoenpflug and P. Richter, Text. Chem. Color. 7 (1975) pp. 13–17.

[169] T.J. Mason, *Sonochemistry: The Use of Ultrasound in Chemistry*, Royal Society of Chemistry, Cambridge, UK, 1990.

[170] C.G. Jeffrey and R.H. Ottewill, Colloid Polym. Sci. 266 (1988) pp. 173–179.

[171] W.J. Lee and J.P. Kim, J. Kor. Fiber Soc. 35 (1998) pp. 100–104.

[172] J. Skoufis, Text. Chem. Color. 11 (1979) pp. 106–109.

[173] A. Datyner, Rev. Prog. Color. Rel. Top. 23 (1993) pp. 40–50.

[174] M. Gorensek, V. Gaber, N. Peternelj and V. Vrhunc, J. Soc. Dyers Colour. 111 (1995) pp. 19–21.

[175] J.D. Turner and M. Chanin, Am. Dyest. Rep. 51 (1962) pp. 780–785.

[176] J.F. Leuck, Am. Dyest. Rep. 68 (1979) pp. 49–54.

[177] A.H. Brown and A.T. Peters, J. Soc. Dyers Colour. 57 (1968) pp. 281–288.

[178] F.M. Rawicz, D.M. Cates and H.A. Rutherford, Am. Dyest. Rep. 50 (320–323) (1961) p. 354.

[179] F. Schlaeppi, R. Wagner and J. McNeill, Text. Chem. Color. 14 (1982) pp. 29–42.

[180] J. Bone, AATCC Rev. 1 (2001) pp. 19–21.

[181] J. Shore, *Cellulosic Dyeing*, Society of Dyers and Colourists, Alden Press, Oxford, UK, 1995.

[182] J. Shore, *Blends Dyeing*, Society of Dyers and Colourists, Bradford, UK, 1998.

[183] Sandoz, (1984) , Colourage 3, 31–36.

[184] T.S. Cannito, Am. Dyest. Rep. 70(3) (1981) p. 23.

[185] Y. Sato and N. Kayaku, Am. Dyest. Rep. 72 (1983) pp. 30–34.

[186] A. Varga and G. Holczer, Melliand (English ed.) 14 (1985) pp. 820–824.

[187] M.D. Teli, Colourage 10 (1988) pp. 11–14.

[188] J.-H. Choi, S.H. Hong, E.J. Lee and A.D. Towns, Color. Technol. 116 (2000) pp. 278–273.

[189] B.S. Ashkenazi, Colourage 3 (1985) pp. 37–39.

[190] J. Park, *A Practical Introduction to Yarn Dyeing*, Society of Dyers and Colourists, West Yorkshire, UK, 1981.

[191] B. Smith, Am. Dyest. Rep. 10 (1987) pp. 34–57.

[192] E. Trotman, *Dyeing and Chemical Technology of Textile Fibres*, 3rd ed., Charles Griffin, London, 1964.

[193] F. Maldonado, A. Ciurlizza, R. Radillo and E.P.D. León, Color. Technol. 116(11) (2000) pp. 359–362.

[194] M.C. Thiry, AATCC Rev. 8(4) (2008) pp. 22–28.

[195] J. Park, J. Soc. Dyers Colour. 107 (1991) pp. 193–196.

[196] J. Park, Rev. Prog. Color. Rel. Top. 34 (2004) pp. 86–94.

[197] W.S. Shim, *A diagnostic expert system for the coloration of polyester materials*, PhD dissertation. North Carolina State University, Raleigh, NC, 2009.